7/20

PERIODIC TABLE OF THE ELEMENTS

I	II A	III A	IV A	V A	VI A	VII A	VIII	VIII	VIII	I B	II B	III B	IV B	V B	VI B	VII B	O
1·008 **H** 1																	4·003 **He** 2
6·940 **Li** 3	9·012 **Be** 4											10·811 **B** 5	12·011 **C** 6	14·007 **N** 7	15·999 **O** 8	18·998 **F** 9	20·183 **Ne** 10
22·990 **Na** 11	24·312 **Mg** 12											26·982 **Al** 13	28·086 **Si** 14	30·974 **P** 15	32·064 **S** 16	35·453 **Cl** 17	39·948 **A** 18
39·102 **K** 19	40·080 **Ca** 20	44·956 **Sc** 21	47·900 **Ti** 22	50·942 **V** 23	51·996 **Cr** 24	54·938 **Mn** 25	55·847 **Fe** 26	58·993 **Co** 27	58·710 **Ni** 28	63·540 **Cu** 29	65·370 **Zn** 30	69·720 **Ga** 31	72·590 **Ge** 32	74·922 **As** 33	78·960 **Se** 34	79·909 **Br** 35	83·800 **Kr** 36
85·470 **Rb** 37	87·620 **Sr** 38	88·905 **Y** 39	91·220 **Zr** 40	92·906 **Nb** 41	95·940 **Mo** 42	99 **Tc** 43	101·070 **Ru** 44	102·905 **Rh** 45	106·400 **Pd** 46	107·870 **Ag** 47	112·400 **Cd** 48	114·820 **In** 49	118·690 **Sn** 50	121·750 **Sb** 51	127·600 **Te** 52	126·904 **I** 53	131·300 **Xe** 54
132·905 **Cs** 55	137·340 **Ba** 56	RARE EARTHS **La-Lu**	178·490 **Hf** 72	180·948 **Ta** 73	183·85 **W** 74	186·20 **Re** 75	190·2 **Os** 76	192·2 **Ir** 77	195·09 **Pt** 78	196·967 **Au** 79	200·590 **Hg** 80	204·370 **Tl** 81	207·190 **Pb** 82	208·980 **Bi** 83	210 **Po** 84	210 **At** 85	222 **Rn** 86
223 **Fr** 87	227 **Ra** 88	ACTINIDE SERIES **Ac-Lw**															

RARE EARTHS	138·92 **La** 57	140·13 **Ce** 58	140·92 **Pr** 59	144·24 **Nd** 60	145 **Pm** 61	150·35 **Sm** 62	152·0 **Eu** 63	157·25 **Gd** 64	158·92 **Tb** 65	162·50 **Dy** 66	164·93 **Ho** 67	167·3 **Er** 68	169 **Tm** 69	173·04 **Yb** 70	174·98 **Lu** 71
ACTINIDE SERIES	227 **Ac** 89	232·038 **Th** 90	231 **Pa** 91	238·030 **U** 92	237 **Np** 93	242 **Pu** 94	243 **Am** 95	247 **Cm** 96	249 **Bk** 97	251 **Cf** 98	254 **E** 99	255 **Fm** 100	256 **Mv** 101	255 **No** 102	257 **Lw** 103

The atomic number of an element appears below its symbol, while its atomic weight is above.

Materials Science

Materials Science

J. C. ANDERSON
Imperial College of Science and Technology

K. D. LEAVER
Imperial College of Science and Technology

Nelson

THOMAS NELSON AND SONS LTD
36 Park Street London W1
P.O. Box 336 Apapa Lagos
P.O. Box 25012 Nairobi
P.O. Box 21149 Dar es Salaam
P.O. Box 2187 Accra
77 Coffee Street San Fernando Trinidad

THOMAS NELSON (AUSTRALIA) LTD
597 Little Collins Street Melbourne 3000

THOMAS NELSON AND SONS (SOUTH AFRICA) (PROPRIETARY) LTD
51 Commissioner Street Johannesburg

THOMAS NELSON AND SONS (CANADA) LTD
81 Curlew Drive Don Mills Ontario

Illustrations by Colin Rattray

First published in Great Britain 1969

17 761007 7 (Boards)
17 771009 8 (Paper)

Printed in Great Britain by
Butler & Tanner Ltd, Frome and London

Contents

Contents

Contents

Contents

Preface

The study of the science of materials has become in recent years an integral part of virtually all university courses in engineering. The physicist, the chemist, and the metallurgist may, rightly, claim that they study materials scientifically, but the reason for the emergence of the 'new' subject of materials science is that it encompasses all these disciplines. It was with this in mind that the present book was written. We hope that, in addition to providing for the engineer an introductory text on the structure and properties of engineering materials, the book will assist the the student of physics, chemistry, or metallurgy to comprehend the essential unity of these subjects under the all-embracing, though ill-defined, title 'Materials science'.

The text is based on the introductory materials course given to all engineers at Imperial College, London. One of the problems in teaching an introductory course arises from the varying amounts of background material possessed by the students. We have, therefore, assumed only an elementary knowledge of chemistry and a reasonable grounding in physics, since this is the combination most frequently encountered in engineering faculties. On the other hand, the student with a good grasp of more advanced chemistry will not find the treatment familiar and therefore dull. This is because of the novel approach to the teaching of basic atomic structure, in which the ideas of wave mechanics are used, in a simplified form, from the outset. We believe that this method has several virtues: not only does it provide for a smooth development of the electronic properties of materials, but it inculcates a feeling for the uncertainty principle and for thinking in terms of probability, which are more fundamental than the deterministic picture of a particle electron moving along a specific orbit about the nucleus. We recognize that this approach is conceptually difficult, but no more so than the conventional one if one remembers the 'act of faith' which is necessary to accept the quantization condition in the Bohr theory. The success of this approach with our own students reinforces the belief that this is the right way to begin.

In view of the differences which are bound to exist between courses given in different universities and colleges, some of the more advanced material has been separated from the main body of the text and placed at the end of the appropriate chapter. These sections, which are marked with an asterisk, may, therefore, be omitted by the reader without impairing comprehension of later chapters.

In writing a book of this kind, one accumulates indebtedness to a wide range of people, not least to the authors of earlier books in the field. We particularly wish to acknowledge the help and encouragement given by our academic colleagues, especially J. M. Alexander and F. Ellis whose careful reading and advice on the chapters covering mechanical properties was much appreciated.

Our students have given us much welcome stimulation and the direct help of many of our graduate students is gratefully acknowledged. Finally, we wish to express our thanks to the publishers, who have been a constant source of encouragement and assistance, and to Miss R. Dubinski and Mrs. J. Jackson who, with great patience, typed the manuscript.

J.C.A.
K.D.L.

1 Building blocks: the electron

1·1 Introduction

Science is very much concerned with the identification of patterns, and the recognition of these patterns is the first step in a process that leads to identification of the building bricks with which the patterns are constructed. This process has all the challenge and excitement of exploration combined with the fascination of a good detective story and it lies at the heart of materials science.

At the end of the nineteenth century a pattern of chemical properties of elements had begun to emerge and this was fully recognized by Mendeléev when he constructed his periodic table. Immediately it was apparent that there must be common properties and similar types of behaviour among the atoms of the different elements and the long process of understanding atomic structure had begun. There were many wonders along the way. For instance, was it not remarkable that *only* iron, nickel, and cobalt showed the property of ferromagnetism? (Gadolinium was a fourth ferromagnetic element discovered later.) A satisfactory theory of the atom must be able to explain this apparent oddity. Not only was magnetism exclusive to these elements, but actual pieces of the materials sometimes appear magnetized and sometimes do not, depending on their history. Thus a theory that merely states that the atoms of the element are magnetic is not enough; we must consider what happens when the atoms come together to form a solid.

Similarly, we wonder at the extraordinary range—and beauty—of the shapes of crystals; here we have patterns—how can they be explained? Why are metals ductile while rocks are brittle? What rules determine the strength of a material and is there a theoretical limit to the strength? Why do metals conduct electricity while insulators do not? All these questions are to do with the properties of aggregations of atoms. Thus an understanding of the atom must be followed by an understanding of how atoms interact when they form a solid because this must be the foundation on which explanations of the properties of materials are based.

This book attempts to describe the modern theories through which many of these questions have been answered. We are not interested in tracing their history of development but prefer to present, from the beginning, the quantum-mechanical concepts that have been so successful in modern atomic theory. The pattern of the book parallels the pattern of understanding outlined above. We must start with a thorough grasp of fundamental atomic theory and go on to the theories of aggregations of atoms. As each theoretical concept emerges it is used to explain relevant observed properties. With such a foundation the electrical, mechanical, thermal, and other properties of materials can be described, discussed, and explained.

To begin at the beginning we consider the building bricks of the atoms themselves, starting with the electron.

1·2 The electron

The electron was first clearly identified as an elementary particle by J. J. Thomson in 1897. A more detailed description of his experiment is given towards the end of this chapter; here it is sufficient to say that he was able to conclude that the electron is a constituent of all matter. It was shown to have a fixed, negative charge, e, of $1 \cdot 6021 \times 10^{-19}$ coulomb and a mass of $9 \cdot 1085 \times 10^{-31}$ kilogramme at rest. Thomson's proof of the existence of the electron was the essential prerequisite for the subsequent theories of the structure of the atom. However, before going on to consider these we must first review some of the known facts about atoms themselves.

1·3 Avogadro's number

In the electrolysis of water, in which a voltage is applied between two electrodes immersed in the water, hydrogen is observed to be given off at the negative electrode (the cathode). This indicates that the hydrogen ion carries a positive charge and measurement shows that it takes 95,650 coulombs of electricity to liberate one gramme of hydrogen. Now if we know how many atoms of hydrogen make up one gramme it would be possible to calculate the charge per hydrogen ion. This can be done using Avogadro's number, but before defining it we must describe what is meant by a 'gramme-atom' and a 'gramme-molecule' (or 'mole').

It is known from chemistry that all substances are either elements or compounds and that compounds are made up of elements. Any quantity of an element is assumed to be made up of atoms, all of equal size and mass, and the mass of each atom when expressed in terms of the mass of the hydrogen atom was defined as its atomic weight. In 1815 Prout suggested that if the atomic weight of hydrogen were taken as unity the atomic weights of all other elements should be whole numbers. This turned out not to be quite true and internationally agreed atomic weights were based, instead, on the atomic weight of oxygen being 16, which gives hydrogen the atomic weight 1·0080.

It was later discovered that an element could have different isotopes, i.e., atoms of the same atomic *number* could have differing atomic *weights*: this will be discussed in a later chapter. Mixtures of naturally occurring isotopes were the cause of the atomic weights not being exactly whole numbers. In 1962 it was internationally agreed to use the isotope C^{12}, with an atomic weight of 12, as the basis of all atomic weights, which still gives to hydrogen the value 1·0080.

Thus a gramme-atom of a substance is the amount of substance whose mass in grammes equals its atomic weight. Similarly a gramme-molecule (or mole) of a compound is the amount whose mass in grammes equals its molecular weight which, in turn, is the sum of the atomic weights of the atoms which go to make up the molecule.

We may now define Avogadro's number, which is the number of atoms in a gramme-atom (or molecules in a gramme-molecule) of any substance and it is a universal constant.

Returning to our electrolysis experiment, the amount of electricity required to liberate a gramme-atom of hydrogen will be $95,650 \times 1·0080 = 96,420$ coulombs. Now, suppose we *assume* that the charge on the hydrogen ion is equal to that on the electron, then we may calculate Avogadro's number, N.

$$N = \frac{96,420}{1·602 \times 10^{-19}} = 6·02 \times 10^{23} \text{ mole}^{-1}$$

Experimentally, the accurate value has been determined as $6·023 \times 10^{23}$ and so it may be concluded that the charge on the hydrogen ion is equal in magnitude and opposite in sign to that on the electron.

Also using Avogadro's number we may calculate the mass of a hydrogen atom as

$$\frac{\text{Atomic weight}}{\text{Avogadro's number}} = \frac{1·0080}{6·023 \times 10^{23}}$$

$$= 1·672 \times 10^{-24} \text{ gramme}$$

This is just 1,840 times the mass of the electron.

Such a calculation suggests that there is another constituent of atoms, apart from the electron, which is relatively much more massive and which is positively charged.

1·4 The Rutherford atom

In 1911 Rutherford made use of the a-particle emission from a radioactive source to make the first exploration of the structure of the atom. By passing a stream of particles through a thin gold foil and measuring the angles through which the beam of particles was scattered he was able to conclude that most of the mass of the gold atom (atomic weight 187) resided in a small volume called the nucleus which carried a positive charge. He was also able to show that the radius of the gold nucleus is not greater than 3.2×10^{-12} cm.

Since the number of gold atoms in a gramme-molecule of gold is given by Avogadro's number and, from the density (mass per unit volume) we may calculate the volume occupied by a gramme-molecule, then we may make an estimate of the volume of each atom. If we assume them to be spheres packed together as closely as possible we can deduce that the radius of the gold atom is in the region of 10^{-8} cm, which is 10,000 times larger than the radius obtained by Rutherford. Thus the atom seems to comprise mainly empty space.

To incorporate these findings and the discovery of the electron Rutherford proposed a 'planetary' model of the atom; this postulated a small dense nucleus carrying a positive charge about which orbited the negative electrons like planets round the sun. The positive charge on the nucleus was taken to be equal to the sum of the electron charges so that the atom was electrically neutral.

This proposal has an attractive simplicity. Referring to Fig. 1·1: if the electron moves in a circular orbit of radius, r, with a constant linear velocity, v, then it will be subject to two forces. Acting inwards will be the force of electrostatic attraction described by Coulomb's law:

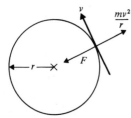

$$F = \frac{q_1 q_2}{4\pi\varepsilon_0 r^2} \tag{1·1}$$

Fig. 1·1 Rutherford's planetary atomic model

where q_1 and q_2 are the positive and negative charges, in this case each having the value of the charge, e, on the electron and ε_0 is the permittivity of free space given by $10^{-9}/36\pi$.

Acting outwards there will be the usual centrifugal force given by mv^2/r, where m is the electron mass and v^2/r is its radial acceleration. The orbit of the electron will settle down to a stable value where these two forces just balance each other, that is, when

$$\frac{mv^2}{r} = \frac{e^2}{4\pi\varepsilon_0 r^2} \tag{1·2}$$

Unfortunately, this model has a basic flaw. Maxwell's equations, which describe the laws of electromagnetic radiation, can be used to show that an electron undergoing a change in velocity (i.e., acceleration and deceleration) will radiate energy in the form of electromagnetic waves. This can be seen by analogy: if the electron in a circular orbit were viewed from the side it would appear to be travelling rapidly backwards and forwards. Now it carries a charge and a moving charge represents an electric current. Thus its alternating motion corresponds to an alternating electric current at a very high frequency. Just such a high-frequency alternating current is supplied to an aerial by a radio transmitter and the electromagnetic radiation from the aerial is readily detectable. Thus we must expect the electron in Rutherford's model of the atom continuously to radiate energy. More formally, since an electron moving in an orbit of radius, r, with a linear velocity, v, is, by the laws of Newtonian mechanics, subject to a continual radial acceleration of magnitude v^2/r, it must, by the laws of electromagnetic radiation, continuously radiate energy.

The kinetic energy of the electron in its orbit is proportional to the square of its velocity and if it loses energy its velocity must diminish. The centrifugal force, mv^2/r, due to the radial component of acceleration will therefore also diminish and so the electrostatic attraction between the positive nucleus and the negative electron will pull the electron closer to the nucleus. It is not difficult to see that the electron would ultimately spiral into the nucleus.

The resolution of this difficulty was only made possible by the bringing together of a variety of observations and theories and we must now consider these briefly in turn.

1·5   Waves and particles

One of the more remarkable puzzles which gradually emerged from experimental physics as more and more physical phenomena were explored was that it appeared that light, which was normally regarded as a wave (an electromagnetic wave to be precise), sometimes behaved as if the ray of light were a stream of particles. Similarly, experimental evidence emerged suggesting that electrons, which we have so far treated as particles, may behave like waves. If this is so, the Rutherford atom, based on treating the electron as a particle having a fixed mass and charge and obeying Newtonian mechanics, is evidently too crude a model. This is obviously a crucial matter and we must examine the relevant experiments carefully.

a. Electron waves

Remember that the properties which convince us that light is a wave motion are *diffraction* and *interference*. For instance, a beam of light impinging on a very narrow slit is diffracted (i.e., spreads out behind the slit). A beam of light shining on two narrow adjacent slits, as shown in Fig. 1·2, is diffracted and if a screen is placed beyond the slits interference occurs, giving a characteristic intensity distribution as shown. Yet another pattern is produced if we use many parallel slits, the device so

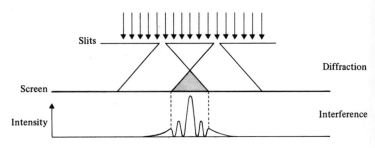

Fig. 1·2 Diffraction of light through two slits

produced being called a *diffraction grating*. As shown in Fig. 1·3, if we take a line source of light and observe the image it produces when viewed through a diffraction grating we see a set of lines whose separation depends on the wavelength of the light and the distance between the slits in the diffraction grating. If two diffraction gratings are superimposed with their lines intersecting at right-angles a spot pattern is produced in which the distances between the spots are a function of the wavelength of the light for a given pair of gratings. For the experiment to work the dimensions of the slits in the grating *must be comparable to the wavelengths of the light*, i.e., the slits must be about 10^{-4} cm wide and separated by a similar distance.

If we are to observe the wavelike behaviour of electrons, we can expect to do so with a diffraction grating of suitable dimensions; but we do not yet know the dimensions necessary. It so happens that the size required is very small indeed and, in fact, is comparable with the mean distance between atoms in a typical crystal, that is, in the region of 10^{-8} cm. Thus if we 'shine' a beam of electrons on a crystal which is thin enough for them to pass through, the crystal will act as a diffraction grating if the electrons behave as waves.

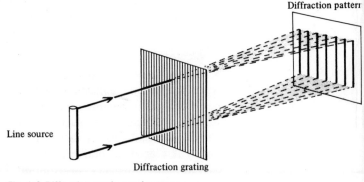

Fig. 1·3 Diffraction grating and pattern

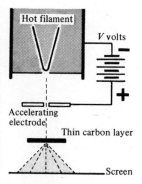

Fig. 1·4 Schematic arrangement to demonstrate electron diffraction

Referring to Fig. 1·4, a source of electrons, such as a hot filament, is mounted in an evacuated enclosure with suitable electrodes such that the electrons can be accelerated and collimated into a narrow beam. The beam is allowed to fall upon a thin carbon layer on the other side of which is placed a fluorescent screen of the type commonly used in television tubes. Electrons arriving at the screen produce a glow of light proportional to their number and energy.

The energy imparted to the electron beam is determined by the potential, V, applied to the accelerating electrode. If we are regarding it as a charged particle the potential energy of an electron when the applied voltage is V volts is, by definition of potential, just eV, where e is its charge in coulombs. The accelerating electron increases its kinetic energy at the expense of its potential energy, and, if we treat it as a particle, it will reach a velocity, v, given by:

$$\tfrac{1}{2}mv^2 = eV \tag{1·3}$$

This is a relationship of which we shall make much use later.

If the electrons are to be regarded as waves it is not clear how we can attribute a kinetic energy of $\tfrac{1}{2}mv^2$ to them since we do not normally attribute mass to a wave. We must just say that they have gained energy and it is convenient to define this energy as the change in potential energy of the wave in passing through a potential difference of V volts. This energy will be eV joules and we therefore define a new unit of energy, the *electron volt*. This is defined as the energy which an electron acquires when it falls through a potential drop of one volt and is equal to $1·602 \times 10^{-19}$ joule.

Returning to Fig. 1·4, the electrons thus impinge upon the carbon layer with high energy. If they were particles we would expect them to be scattered randomly in all directions when they hit atoms in the carbon crystals. The intensity distribution beyond the carbon layer would then be greatest at the centre, falling off uniformly towards the edges, as shown in Fig. 1·5, and the pattern of light on the screen would be a diffuse one, shading off gradually in brightness from the centre outwards.

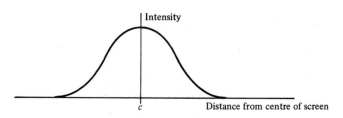

Fig. 1·5 Intensity distribution for randomly scattered light

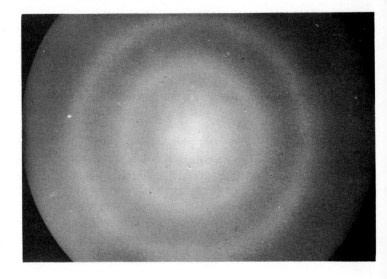

Fig. 1·6 Electron diffraction rings

In fact the experiment produces a quite different result. The pattern seen on the screen consists of rings of intense light separated by regions in which no electrons are detected, as shown in Fig. 1·6.

This can only be explained by assuming that the electron is a wave. If layers of atoms in a crystal act as a set of diffraction gratings we can regard them as line gratings superimposed with their slits at right-angles and the diffraction pattern should be a set of spots, as described earlier. The pattern seen is a series of rings rather than spots because the carbon layer is made up of many tiny crystals all orientated in different directions; the rings correspond to many spot patterns, with differing orientations, superimposed. Each ring corresponds to diffraction from a different set of atomic planes in the crystal. Diffraction of electrons by crystals is a powerful tool in the determination of crystal structures and is dealt with in detail in more advanced textbooks (see, for instance, Azaroff and Brophy).

If in the experiment the accelerating voltage, V, is varied, we observe that the diameters of the rings change, and it is not difficult to establish that the radius, R, of any ring is inversely proportional to the square root of the accelerating voltage. That is,

$$R \propto \frac{1}{V^{1/2}}$$

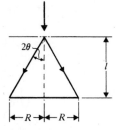

Fig. 1·7 Illustrating Bragg's law of diffraction

To understand this we must go to yet another branch of physics —that of X-ray crystallography. X-rays are electromagnetic waves having the same nature as that of light but with much shorter wavelengths. 'Hard' X-rays have wavelengths in the range of 1 to 10 Å (one ångstrom unit, denoted by 1 Å, is 10^{-10} metres) and are therefore diffracted by the planes of atoms in a crystal, in the same way as light is diffracted by a diffraction grating. Study of X-ray diffraction by crystals led to the well-known Bragg's law relating wavelength to the angle through which a ray is diffracted. This is derived in Chapter 6 and the law can be written as

$$n\lambda = 2d \sin \theta \qquad (1·4)$$

when n is an integer (the 'order' of the diffracted ray), λ the wavelength, d the distance between the planes of atoms, and 2θ is the angle between the direction of the diffracted ray and that of the emergent ray as shown in Fig. 1·7. Since the electrons appear to be behaving as waves it is reasonable to assume that we may also apply Bragg's law to them.

Referring to Fig. 1·7, in our experiment the electron beam fell normally on the carbon film and, on emergence, had been split into diffracted rays forming a cone of apical angle 4θ. From the diagram, the radius, R, of the diffraction ring is given by

$$R = l \tan 2\theta \qquad (1·5)$$

where l is the distance from the carbon film to the screen. Since θ is commonly only a few degrees we may approximate $\tan 2\theta \approx 2\theta \approx 2 \sin \theta$, so that

$$R \approx 2l \sin \theta$$

and, using Bragg's law of Eqn (1·4),

$$R \approx \frac{nl\lambda}{d} \qquad (1·6)$$

Since R and l can be measured and n is known for the 'order' of the ring ($n = 1$ for the first ring), we may determine λ from this experiment, given a knowledge of the spacing, d, between the atomic planes in carbon, which is known from X-ray diffraction work.

The importance of this result for our present purposes is that it shows R is proportional to λ, the other quantities being constants. But we have already observed that $R \propto 1/V^{1/2}$ and so we have the result that

$$\lambda \propto \frac{1}{V^{1/2}}$$

Now we earlier defined the energy of the wave, E, as the potential through which it fell, that is

$$E = eV \tag{1·7}$$

Since e is a constant it follows that

$$\lambda \propto \frac{1}{E^{1/2}} \tag{1·8}$$

and we have the important result that the electron wavelength is inversely proportional to the square root of the energy of the wave.

Now although the electron might be behaving like a wave we do know that it can also behave as if it were a particle whose energy is given by

$$E = \tfrac{1}{2}mv^2 = eV$$

and which will have a momentum, p, given by $p = mv$. Relating the kinetic energy, E, to the momentum we have

$$E = \tfrac{1}{2}mv^2 = \frac{p^2}{2m} \tag{1·9}$$

Thus, since m is a constant, $E^{1/2} \propto p$ and from the experimental result of Eqn (1·8), we may write

$$\lambda \propto \frac{1}{p} \tag{1·10}$$

Thus we are postulating that the electron wave carries a momentum which is inversely proportional to its wavelength. If the wave is to carry momentum we are implying that the wave, in some way, represents a mass, m, moving with a velocity, v, and we may refer to the electron as a 'matter wave'.

The relationship between λ and p was, in fact, proposed by de Broglie before the phenomenon of electron diffraction was discovered. Putting in a constant of proportionality, h, we obtain de Broglie's relationship,

$$\lambda = \frac{h}{p}$$

in which h is a universal constant called *Planck's constant* and has the value $6·625 \times 10^{-34}$ Nm s ($=$ joule second). This is one of the most important relationships in physics and applies to all matter, from electrons to billiard balls. It implies that all moving matter will exhibit wavelike behaviour. However, because of the small value of Planck's constant, the wavelength of anything much larger than an electron will be so short that we will never see any evidence, such as diffraction, of wavelike behaviour. (See problems at the end of this chapter.)

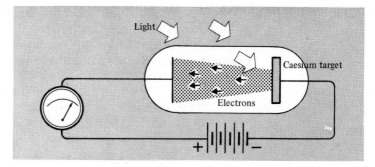

Fig. 1·8 Schematic arrangement to demonstrate photo-emission

b. Particles of Light

We have already mentioned that the properties of diffraction and interference show that light is a wave motion. But now consider the experiment shown in Fig. 1·8. Some materials (e.g. caesium) when irradiated with light will emit electrons which can be detected by attracting them to a positively charged electrode and observing the flow of current.

It can be shown from other experiments that to remove an electron from a solid it is necessary to give the electron enough kinetic energy. 'Enough' here means more than some minimum quantity, W, which represents the 'potential energy barrier' (often called just the 'potential barrier') which the electron has to surmount in order to leave the solid. Energy in excess of this amount, W, supplied by the light beam must reappear as kinetic energy, E_k, of the electron, i.e.,

$$\text{Total energy supplied by light} = E_L = W + E_k \qquad (1\cdot11)$$

If we reduce E_L by reducing the amount of light supplied E_k must also decrease. So reducing the *intensity* of the light should reduce E_k which is measurable.

But this does not happen! What we observe is that the *number* of electrons released by the light reduces proportionally with the light intensity but the *energy* of the electrons remains the same.

This can be explained by supposing the light beam to consist of particles of fixed energy. Reducing the intensity would reduce the number of particles flowing per unit time. If we suppose that when a particle 'collides' with an electron it can give the electron its whole energy, then reducing the rate of flow of particles would reduce the number of collisions but the energy transferred at each collision would remain the same. Thus we would reduce the rate of emission of electrons without changing their kinetic energy.

We call the light particles 'photons'. How do we know how much energy a photon has?

We can get one clue by changing the *frequency* of the light. We then find that the emitted electrons have a different energy. By trying several different wavelengths (frequencies) it is possible to show that the energy, E, of each photon is proportional to the frequency, f, that is,

$$E \propto f$$

This is clearly different from the case of the electron for which $E^{1/2} \propto f$ [see Eqn (1·8)]. But the constant of proportionality is again h, Planck's constant, so that

$$E = hf \tag{1·12}$$

The implications of this are profound since we are saying that the light waves behave as if they consist of a stream of particles, each particle carrying an energy hf. Thus the energy carried by the wave is apparently broken up into discrete lumps or 'quanta'. The energy of each quantum depends only on the frequency (or wavelength) of the light while the intensity (or brightness) of the light is determined by the number of quanta arriving on the illuminated surface in each second. This idea emerged just about the time of the discovery of the electron. It was in 1900 that Planck suggested that heat radiation, which is infra-red light, is emitted or absorbed in multiples of a definite amount. This definite amount was named a quantum and its energy was defined as hf, where h was a universal constant that came to be called Planck's constant. Just as wavelike behaviour can be assumed to apply to all moving matter, so all energy associated with it is *quantized* whatever its form. In most practical applications in the real world, such as in the oscillations of a pendulum, the frequency is so low that single quantum could never be observed. However, at the high frequencies of light, X-rays, and electron waves, quantum effects will be readily apparent. This general truth is the basis of the quantum theory and led to a new type of mathematics being developed for application to physical problems.

1·6 Particles or waves?

We have seen that both light and electrons behave in some ways like particles, in others like waves. How is this possible? One way of understanding this is to think about the many possible forms a wave may take. The first is a simple plane wave. If we plot amplitude vertically and distance horizontally we can draw such a wave in one dimension as in Fig. 1·9(a). This wave has a single wavelength and frequency. If we add together two plane waves with slightly different wavelengths the result is a single wave of different 'shape' as shown in Fig. 1·9(b). Its amplitude

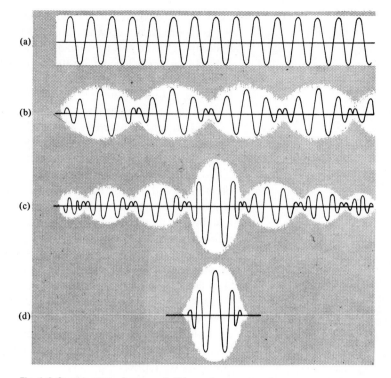

Fig. 1·9 Combination of waves of different frequency to form a wave packet

varies with position so that at regular intervals there is a maximum. Adding a third plane wave with yet another wavelength makes some of these maxima bigger than others. If enough plane waves are added together just *one* of the maxima is bigger than all the others as in Fig. 1·9(c): if they are added together in the correct phases all the other maxima disappear and the wave so formed has a noticeable amplitude only in one (relatively) small region of space as shown in Fig. 1·9(d). This type of wave is often called a *wave packet* and it clearly has some resemblance to what we call a particle, in that it is localized.

So we can regard an electron or a light wave when it acts in the manner of what we call a 'particle', as being a wave packet. It is still useful at such times to use the language which applies to particles, just as a matter of convenience.

1·7   Wave nature of larger 'particles'

We have seen that to treat the electron as a 'particle' is an approximate way of describing a wave packet. The wavelike behaviour shows up only when the electron wavelength is comparable, for example, to the diameter of an atom or the distance between atoms. This is the case, too, with light; its wave nature shows up when it passes through a slit whose width is comparable to the wavelength of the light.

Now an atom, being matter like the electron, also has a characteristic wavelength given by the relationship $\lambda = h/p$. It will thus show wavelike properties when it interacts with something of the right size: roughly, equal to its wavelength.

But the lightest atom is about 2,000 times heavier than the electron, so for the same velocity its momentum, p, is about 2,000 times greater and so the wavelength, λ, is correspondingly 2,000 times smaller than that of the electron. Thus its wavelike properties will almost never be observable since its wavelength is so small. The same applies, even more so, as we consider bigger and bigger 'particles'. So a dust 'particle' consisting of several million atoms virtually never displays wavelike behaviour and can always be treated as a particle. This means, in terms of physics, that Newton's laws will apply to a dust particle while they cannot be expected to apply, at any rate in the same form, to electrons. Thus we see that the difficulty in the Rutherford atom lies in treating the electron as a particle subject to Newtonian mechanics. It is necessary to develop a new mechanics—wave mechanics—to deal with the problem and this is the subject of the next chapter.

*1·8   The Thomson experiment

The following descriptions are not historically accurate but, rather, are intended to bring out the basic behaviour of electrons in the presence of electric and magnetic fields.

In the remainder of this chapter we deal in more detail with some of the ground covered earlier; these sections may be omitted on first reading.

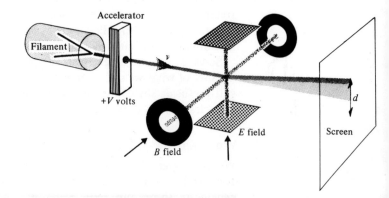

Fig. 1·10 Schematic diagram of Thomson's experiment

Beginning with Thomson's type of experiment, suppose that we have the apparatus illustrated in Fig. 1·10. Electrons are emitted from a hot filament, are focused into a beam, and accelerated by a potential V to a velocity v, finally impinging on a fluorescent screen. The beam is arranged to pass between two plane parallel electrodes between which a voltage can be applied to produce a uniform electrostatic field, E volt/m, across the path of the beam. If the electrons are treated as charged particles they will be deflected by the field by virtue of the force F_ε newtons perpendicular to their path which it exerts on them. This force is given, from the definition of electric field, by

$$F_\varepsilon = eE$$

and the resulting deflection of the beam moves its arrival point at the screen through a distance, d.

In the same position as the electrodes we also have a pair of magnetic coils which, when supplied with current, produce a uniform magnetic field of flux density B Wb/m² (or teslas) perpendicular to the path of the electron beam. Now the moving electrons, since they are charged, are equivalent to a current of value ev/l over a path of length, l. One of the fundamental laws of electromagnetism is that a wire carrying a current, i, in a uniform magnetic flux density, B, experiences a force perpendicular to both the direction of current and of the magnetic field given by Bil, where l is the length of the wire. In the present case, the electrons will therefore experience a force given by

$$F_{\mathrm{H}} = Bev \tag{1·13}$$

If now we arrange that the forces due to the electric and magnetic fields oppose each other and we adjust the fields so that the spot on the screen is returned to its original straight-through position we have

$$eE = Bev \tag{1·14}$$

Now the velocity of the beam, as we have seen previously, is given by the relationship

$$eV = \tfrac{1}{2}mv^2 \tag{1·15}$$

so that

$$v = \left(\frac{2eV}{m}\right)^{1/2}$$

Thus, substituting in Eqn (1·14),

$$E = B\left(\frac{2eV}{m}\right)^{1/2} \quad \text{or} \quad \frac{e}{m} = \frac{E^2}{2VB^2} \tag{1·16}$$

As E, V, and B are known we have obtained a value of e/m for the electron.

By using different materials for the filament wire it can be established that the same value for e/m is always obtained, thus supporting the statement that the electron is a constituent of all matter.

*1·9 Millikan's experiment

Since the Thomson experiment is only able to yield the charge-to-mass ratio of the electron it is necessary to devise means for measuring either the charge or the mass separately. Thomson designed a suitable method, which was improved in Millikan's famous oil-drop experiment, which provided a value for the charge e. Both methods depend upon the speed with which a small drop of liquid falls in air under the influence of gravity. A particle falling freely under the influence of gravity will reach a velocity, v, after falling through a distance, h, given by $v^2 = 2gh$. However, if the particle is falling in air, the friction, or viscosity, of the air comes into play as the velocity increases and the particle finally reaches a constant terminal velocity. If the particle or drop is spherical with a radius a, this velocity is given by Stokes' law as

$$v = \frac{2ga^2\rho}{9\eta} \qquad (1·17)$$

where ρ is the density of the sphere and η the coefficient of viscosity of the air.

If the drops are sufficiently small, v may be only a fraction of a millimetre per second and may be measured by direct observation. It is then possible to deduce the radius, a, and hence the mass of the drop. This is the prerequisite for the oil-drop experiment.

A mist of oil is produced by spraying through a fine nozzle into a glass chamber. The drops may be observed through the side of the chamber by means of a suitable microscope with a graticule in the eyepiece. The free fall of a suitable drop is observed in order to calculate its mass as outlined above.

At the top and bottom of the chamber are plane parallel electrodes to which a potential can be applied producing a uniform electric field in the vertical direction. The drop of oil is charged by irradiating it with ultra-violet light which causes the loss of an electron so that the drop acquires a positive charge, $+e$. The potential is adjusted so that the drop appears to be stationary when the upward force due to the field (top electrode negative) just balances the downward force due to the weight of the drop. Thus,

$$\frac{eV}{d} = mg \qquad (1 \cdot 18)$$

where V is the potential, and d the distance between the electrodes, m is the mass of the drop, and g the acceleration due to gravity.

The charging of the drop by exciting an electron out of it will be better understood when we have dealt with atomic structure in more detail. It is not possible in this experiment to be sure that the drop has not lost two, three, or even more electrons. However, by observing a large number of drops the charge always works out to be an integral multiple of $1 \cdot 6 \times 10^{-19}$ coulomb. Thus the *smallest* charge the drop can acquire has this value, which we therefore assume to be the charge carried by a single electron. Since there is no *a priori* proof that the drop may charge only by loss of an electron we are, in fact, somewhat jumping to conclusions. However, as will be seen later from the detailed atomic model, the only conceivable explanation is precisely the one proposed.

*1·10 Wave analysis

Earlier in the chapter we referred to the fact that combining sinusoidal waveforms of different frequencies produces a wave whose amplitude, as a function of position, differs from the amplitudes of the two constituent waves. By use of simple trigonometrical formulae it is possible to demonstrate this.

Mathematically, a plane wave whose amplitude varies with time, t, can be represented by the expression

$$y = A \sin \omega t \qquad (1 \cdot 19)$$

where y is the amplitude at any instant of time, A is the maximum amplitude, and ω is 2π times the frequency.

If the wave is travelling in a given direction at a velocity c cm/s, then it will take a time x/c s to reach a point x cm from its point of origin, i.e., the variation of amplitude at the point x will lag, in time, behind the variation at the origin. Thus the wave will then be represented by

$$y = A \sin \left[2\pi f \left(t - \frac{x}{c} \right) \right]$$

or

$$y = A \sin \left[2\pi \left(ft - \frac{x}{\lambda} \right) \right]$$

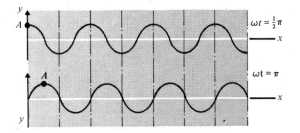

Fig. 1·11 A travelling wave

since $c = f\lambda$; that is

$$y = A \sin\left(\omega t - \frac{2\pi x}{\lambda}\right) \tag{1·20}$$

Using the formula for $\sin(A-B)$ we may write this as

$$y = A\left(\sin \omega t \cos \frac{2\pi x}{\lambda} - \cos \omega t \sin \frac{2\pi x}{\lambda}\right)$$

If we take $\omega t = \pi/2$, then $y = \cos 2\pi x/\lambda$, while, at some sub-sequent time such that $\omega t = \pi$, we have $y = \sin 2\pi x/\lambda$. These are illustrated in Fig. 1·11 and it is easy to see that the equation describes a wave travelling steadily through space.

If we were to combine two waves of the same frequency and amplitude travelling in opposite directions, one in the $+x$ and the other in the $-x$ directions, the individual waves would be given by

$$y_1 = A \sin\left(\omega t - \frac{2\pi x}{\lambda}\right) \quad \text{and} \quad y_2 = A \sin\left(\omega t + \frac{2\pi x}{\lambda}\right)$$

so that using the trigonometrical formulae

$$y = y_1 + y_2 = 2A \sin \omega t \cos \frac{2\pi x}{\lambda} \tag{1·21}$$

In this case the variation of the wave with distance, x, is the same for all times, t, being represented by $\cos 2\pi x/\lambda$ in every case. This is shown in Fig. 1·12 and is called a standing wave. In it there are fixed positions in space, distance $\lambda/2$ apart, where the amplitude of the wave is always zero (nodes) and positions where the wave varies, with time, between zero and a maximum value $2A$ (antinodes).

When we wish to deduce the shape of a combination of travelling waves of different frequencies we may do so by representing them by trigonometric functions having differing values of ω, λ, and amplitude A and adding them up. In principle this can be done by using the trigonometric formulae to simplify the expressions and plotting a graph of the resulting equation.

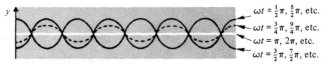

$\omega t = \frac{1}{2}\pi, \frac{5}{2}\pi, \text{etc.}$
$\omega t = \frac{3}{4}\pi, \frac{9}{4}\pi, \text{etc.}$
$\omega t = \pi, 2\pi, \text{etc.}$
$\omega t = \frac{3}{2}\pi, \frac{7}{2}\pi, \text{etc.}$

Fig. 1·12 A standing wave

However, this rapidly becomes clumsy, algebraically, and the simplest method for the more straightforward cases is to draw graphs representing each of the component waves and add them up, point by point. This is, in effect, what has been done in Fig. 1·9.

There is a mathematical technique for simplifying the algebraic complexities which is known as Fourier analysis. This is based on Fourier's theorem which states that any single-valued periodic function, $y = f(t)$, having a period 2π, may be expressed in the form

$$y = C + A_1 \sin(\omega t + \phi_1) + A_2 \sin(2\omega t + \phi_2)$$
$$+ A_3 \sin(3\omega t + \phi_3) + \dots \tag{1·22}$$

Since
$$A \sin(\omega t + \phi) = A \sin \omega t \cos \phi + A \cos \omega t \sin \phi$$
$$= a \sin \omega t + b \cos \omega t$$

where $a = A \cos \phi$ and $b = A \sin \phi$, this expression may be written as

$$y = c + a_1 \sin \omega t + a_2 \sin 2\omega t + a_3 \sin 3\omega t + \dots$$
$$+ b \cos \omega t + b_2 \cos 2\omega t + b_3 \cos 3\omega t + \dots$$

that is
$$y = c + \sum_n a_n \sin n\omega t + \sum_n b_n \cos(n\omega t) \tag{1·23}$$

In order to use Eqn (1·23) to analyse a complex wave it is necessary to determine the coefficients c, a_1, a_2, a_3, . . . and b_1, b_2, b_3, . . . This is done by multiplying both sides of the equation by a suitable factor and integrating between limits. If the multiplying factor is correctly chosen all terms vanish except those which will give the required coefficient.

The details of this are more properly the province of a mathematical textbook. However, we give here the results, together with an example of their application.

$$c = \frac{1}{2\pi} \int_0^{2\pi} y \, d(\omega t) \tag{1·24}$$

$$a_n = \frac{1}{\pi} \int_0^{2\pi} y \sin n\omega t \, d(\omega t) \tag{1·25}$$

$$b_n = \frac{1}{\pi} \int_0^{2\pi} y \cos n\omega t \, d(\omega t) \tag{1·26}$$

where n is a positive integer.

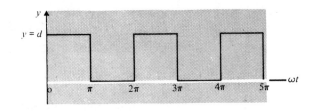

Fig. 1·13 A periodic square wave

It should be realized that for a travelling wave a graph of amplitude against time for a fixed distance will be the same as that for amplitude against distance for a fixed time. In the first instance we are standing at a fixed position and plotting the wave as it goes past and in the second case we take a look at the wave over a length of space at a fixed time; in each case the pattern is the same.

As an example we consider the series of pulses shown in Fig. 1·13 in which amplitude is plotted as a function of ωt. From $\omega t = 0$ to $\omega t = \pi$ the equation for the curve is $y = d$; from $\omega t = \pi$ to $\omega t = 2\pi$ the equation is $y = 0$ and the wave has a period 2π.

Using Eqn (1·24), we write

$$c = \frac{1}{2\pi}\int_0^{2\pi} y\,\mathrm{d}(\omega t) = \frac{1}{2\pi}\int_0^{\pi} y\,\mathrm{d}(\omega t) + \frac{1}{2\pi}\int_\pi^{2\pi} y\,\mathrm{d}(\omega t)$$

$$= \frac{1}{2\pi}\int_0^{\pi} d\,.\,\mathrm{d}(\omega t) + \frac{1}{2\pi}\int_\pi^{2\pi} 0\,.\,\mathrm{d}(\omega t)$$

$$= \frac{1}{2\pi}\,.\,d\pi = \frac{d}{2}$$

This is the mean level of the wave.

Equation (1·25) gives

$$a_n = \frac{1}{\pi}\left[\int_0^{\pi} d\,.\,\sin(n\omega t)\,\mathrm{d}(\omega t) + \int_\pi^{2\pi} 0\,.\,\sin(n\omega t)\,\mathrm{d}(\omega t)\right]$$

$$= \frac{1}{\pi}\left[-\frac{d\cos n\omega t}{n}\right]_0^{\pi}$$

$$= \frac{d}{n\pi}(1 - \cos n\pi)$$

where n is odd the term $(1 - \cos n\pi) = 2$ and when n is even then $(1 - \cos n\pi) = 0$. Thus

$$a_1 = \frac{2d}{\pi} \qquad a_3 = \frac{2d}{3\pi}, \text{ etc.}$$

and

$$a_2 = a_4 = \cdots = 0$$

From Eqn (1·26) we obtain

$$b_n = \frac{1}{\pi}\left[\int_0^\pi d\cos(n\omega t)\,\mathrm{d}(\omega t) + \int_\pi^{2\pi} \cos(n\omega t)\,\mathrm{d}(\omega t)\right]$$

$$= \frac{1}{\pi}\left[\frac{d\sin n\omega t}{n}\right]_0^\pi$$

$$= 0$$

Thus all the cosine terms are zero and the wave is represented by the expression

$$y = \frac{d}{2} + \frac{2d}{\pi}\left(\sin\omega t + \frac{1}{3}\sin 3\omega t + \frac{1}{5}\sin 5\omega t + \ldots\right) \qquad (1·27)$$

Problems

1·1 What reasons are there for believing that matter consists of atoms?

1·2 Outline two experiments, one showing the electron to be a particle, the other showing it to be a wave.

1·3 The proton is a particle with a positive charge equal in magnitude to that of the electron and with a mass 1,840 times that of the electron. Calculate the energy, in electron volts, of a proton with a wavelength of 0·01 Å $(1\text{Å} = 10^{-10}$ m).

1·4 A dust particle of mass 1 μg travels in outer space at a velocity 25,000 mile/h. Calculate its wavelength.

1·5 In an electron diffraction experiment the distance between the carbon film and the screen is 10 cm. The radius of the innermost diffraction ring is 0·35 cm and the accelerating voltage for the electron beam is 10,000 V. Calculate the spacing between the atomic planes of carbon. (3·55 Å)

1·6 X-rays are produced when high-energy electrons are suddenly brought to rest by collision with a solid. If all the energy of each electron is transferred to a photon, calculate the wavelength of the X-rays when 25 keV electrons are used.

1·7 A static charge of 0·1 coulomb is placed by friction on a spherical piece of ebonite of radius 20 cm. How many electronic charges are there per square centimetre of the surface?

1·8 A parallel beam of light of wavelength 0·5 μm has an intensity of 10^{-4} W/cm^2 and is reflected at normal incidence from a mirror. Calculate how many photons are incident on a square centimetre of the mirror in each second.

*1·9 Obtain the Fourier coefficients c, a_n, and b_n, for a periodic wave whose equation is $y = \alpha t$ from 0 to π and $y = 0$ from π to 2π, where α is a constant.

2 Wave mechanics

2·1 Introduction

In the previous chapter we saw that the properties of the electron as a particle (which is localized) and a wave (which extends over a region of space) may be brought together by treating it as a wave packet. We now require to develop a mathematical means of representing it in order that we can perform similar calculations to those for which we use Newtonian mechanics in the case of a particle.

2·2 Matter, waves, and probability

A plane wave travelling in the x direction and having a wavelength, λ, and frequency, f, can be represented by expressions of the type

$$
\left.
\begin{aligned}
\Psi &= A \sin\left(\omega t - \frac{2\pi x}{\lambda}\right) \\
\Psi &= A \cos\left(\omega t - \frac{2\pi x}{\lambda}\right)
\end{aligned}
\right\}
\tag{2·1}
$$

where $t =$ time and $\omega = 2\pi f$. The difference between the sine and cosine functions is simply a shift along the horizontal axis, that is, a difference in phase. Thus we may plot the wave, at a given point in space described by $x = 0$, as shown in Fig. 2·1(a). This corresponds to standing at a particular point and plotting the variation in amplitude of the wave with time as it passes. Alternatively, we may choose to look at a length of space at a given time, described for example by $t = 0$, and plot the variation of amplitude of the wave with position along the length chosen, as in Fig. 2·1(b).

If we describe an electron by such an equation an immediate difficulty arises, namely, what does the amplitude of the wave represent? In all our experiments we can detect only that an electron is present or is not present: we can never detect only a part of it. Thus the electron must be represented by the whole wave and we must somehow relate the point-to-point variation of amplitude to this fact. It was the physicist Born who suggested that the wave should represent the probability of finding the electron at a given point in space and time, that is, Ψ is related to this probability. But Ψ itself cannot actually equal this

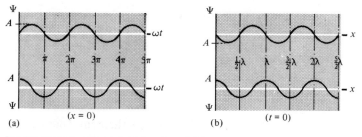

Fig. 2·1 Wave function ψ as a function of (a) ωt for $x = 0$ and (b) position x for $\omega t = 0$

probability since the sine and cosine functions can be negative as well as positive and there can be no such thing as a negative probability. Thus we use the intensity of the wave, which is the square of its amplitude, to define the probability. However, at this point our ideas and our mathematical representation are out of step. We can see that Ψ has no 'real' meaning and yet Eqn (2·1) defines Ψ as a mathematically real quantity. In general, Ψ will be a mathematically complex quantity with real and imaginary parts, that is, we can represent Ψ by a complex number of the type $(A + jB)$ where $j = \sqrt{-1}$.

Now the square of the modulus of a complex number is always a real number and so we can use this to represent the probability defined above. Thus we will have the wave function Ψ in the form

$$\Psi = (A + jB)$$

and the probability given by

$$|\Psi|^2 = (A^2 + B^2).$$

Traditionally—and in most physics textbooks—this is put in a different way which, nevertheless, is precisely the same thing. If $\Psi = (A + jB)$ its complex conjugate is $\Psi^* = (A - jB)$. The probability of finding an electron which is represented by the wave function Ψ at a given point in space and time is defined by the product of Ψ and its complex conjugate Ψ^*, that is:

$$\text{Probability} = \Psi \, \Psi^*$$
$$= (A + jB)(A - jB)$$
$$= (A^2 + B^2)$$
$$= |\Psi|^2$$

It remains to choose a mathematical expression which will represent a complex wave. We can combine a real wave

$$\Psi' = A \cos\left(\omega t + \frac{2\pi x}{\lambda}\right)$$

with an imaginary one

$$\Psi'' = jA \sin\left(\omega t + \frac{2\pi x}{\lambda}\right)$$

to give

$$\Psi = \Psi' + \Psi'' = A\left[\cos\left(\omega t + \frac{2\pi x}{\lambda}\right) + j\sin\left(\omega t + \frac{2\pi x}{\lambda}\right)\right]$$

that is

$$\Psi = A\exp[j(\omega t + kx)] \qquad (2\cdot2)$$

by de Moivre's theorem, where $k = 2\pi/\lambda$.

We can separate the space-dependent and time-dependent parts of Eqn (2·2) by writing it as the product of two exponentials, that is:

$$\Psi = A\exp(\omega t)\exp(jkx) \quad \text{or} \quad \Psi = \psi\exp(j\omega t) \qquad (2\cdot3)$$

where $\psi = \exp(jkx)$ and represents the space variation of the wave. Equation (2·3) is a general expression for a complex wave in which ψ may be any function of x, not necessarily of the form $\exp(jkx)$.

Considering the probability, we have,

$$\begin{aligned}
\Psi\,\Psi^* &= \psi\psi^*\exp(j\omega t)\exp(-j\omega t)\\
&= \psi\psi^*\exp(j\omega t - j\omega t)\\
&= \psi\psi^*\exp(0)\\
&= \psi\psi^*
\end{aligned}$$

Thus if we can find ψ, and hence ψ^*, we can determine the most probable position for the electron in a given region of space. Since the electron carries a charge the function $\psi\psi^*$ gives a measure of the distribution of charge over the same region. These are the most useful applications of the theory and in the majority of cases we can forget about the time-variation term.

2·3 Wave vector, momentum, and energy

The electron has now been described as a wave by means of a mathematical expression. It must be realized that the expression developed describes a wave moving in one direction (the x direction) only. It is perfectly possible to have a wave whose amplitude and wavelength are different in different directions at the same time. For example, we can have a two-dimensional wave as shown in Fig. 2·2. In this the variation of amplitude with distance and the wavelengths are different as one proceeds along the x and along the z directions, and thus the mathematical expression for the amplitude must be a function of x and z as well as of time. Equally, although it is almost impossible to draw, we could clearly have a three-dimensional wave described by a mathematical expression involving three space coordinates. However, even in three dimensions, there will always be a

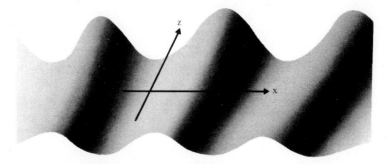

Fig. 2·2 Two dimensional wave

resultant direction of travel for the wave along which its wavelength can be measured.

When we describe a particle in motion we must specify not only its speed but also the direction in which it is moving and this is done by treating its velocity as a vector, **v**. A vector is defined as a quantity having magnitude and direction and the rules for vector summation, multiplication, and so on have been set up in vector algebra.

The student is referred to mathematical textbooks for the details of vector algebra. It is sufficient for our purposes here to remember that a quantity printed in bold type, for example **v**, is a vector. This will have components of v_x, v_y, and v_z in a cartesian coordinate system as shown in Fig. 2·3. The length of the vector is called its scalar magnitude and is printed in ordinary type thus, v, or in italic thus, v. The relationship between the scalar magnitude of the components in a cartesian coordinate system is always given by

$$v = (v_x^2 + v_y^2 + v_z^2)^{1/2}$$

Although the electron is not a particle its momentum has a direction and this is the direction in which the wave itself is travelling. Now the momentum is directly proportioned to $1/\lambda$, the inverse wavelength, and its direction is that along which the wavelength is measured. So we can define a vector, **k**, which has magnitude $2\pi/\lambda$ and the direction defined above and which we call the wave vector. This is, of course, related to the velocity, c, of the wave through the formula, $\lambda = c/f$, but is generally a more convenient quantity to use.

In particular, we can relate the wave vector directly to the momentum, p, carried by the wave through the de Broglie relationship, $p = h/\lambda$. Now, for a particle, $p = mv$ and since v is a vector, p must also be a vector. Similarly with the wave: since $k = 2\pi/\lambda$ we have

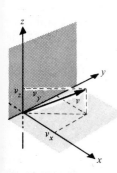

Fig. 2·3 Components of the velocity vector v

$$\mathbf{p} = \frac{h}{2\pi} \cdot \mathbf{k} \qquad (2 \cdot 4)$$

This means that the direction in which momentum is carried by the wave is given by the direction of the wave vector.

In many of our calculations the actual direction of travel of the wave may not be important and k is simply used as an ordinary scalar quantity. This will be the case when the wave is assumed to be a plane wave travelling in a single direction: however, in two-dimensional or three-dimensional problems the vector nature of k must be remembered.

Energy is, of course, a scalar quantity since it can never be said to have a direction. The kinetic energy of a particle is given by $\frac{1}{2}mv^2 = p^2/2m$ where the total velocity or momentum magnitude is used without regard for its direction. Similarly for a wave, the kinetic energy, E_k, associated with it is given by

$$E_k = \frac{p^2}{2m} = \frac{h^2 k^2}{8\pi^2 m} \qquad (2 \cdot 5)$$

[It should be noted that $h/2\pi$ is written conventionally as $\hbar$, (called 'h-bar') and this is often used in textbooks. We then have $p = \hbar k$ and $E_k = \hbar^2 k^2/2m$, but this notation is not used here.]

2·4   Potential energy

By means of the wave vector we can describe the direction of travel of a wave and the energy which it carries. However, we must also be able to describe the environment in which it moves because the electron will, in general, interact with its environment. Since the electron carries a charge a convenient method of doing this is to ascribe to the electron a potential energy, E_p, which may be a function of position. Thus, for example, if the electron, which is negatively charged, is in the vicinity of a positive charge it will have a potential energy in accordance with Coulomb's law of electrostatics. This is proved in textbooks covering elementary electrostatics: it is sufficient for our purposes to define the potential energy of an electron as the work done in bringing an electron from infinity to a point distant, r, from a positive charge of strength, $+e$. This potential energy is given by

$$E_p = \frac{-e^2}{4\pi\varepsilon_0 r} \qquad (2 \cdot 6)$$

where ε_0 is the permittivity of free space.

There will, of course, be other ways in which E_p could vary with position apart from the inverse variation of Eqn (2·2): for

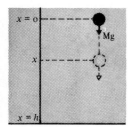

Fig. 2·4 Illustrating
potential energy

example, an electron wave moving through a solid such as a metal encounters positively charged atoms (ions) at regular intervals and in such a case the potential energy is a periodic function of position.

To take a simple mechanical analogy, suppose we have a mass M (kg), at a height h (m) above the floor, as illustrated in Fig. 2·4, and we define the potential energy as zero at floor level. The potential energy of the mass at the initial position will be given by the work done in raising it to that position from the floor, that is, by force × distance.

If the mass is at rest in this position its kinetic energy is zero and its total energy is equal to its potential energy. When it is released it falls, converting its potential energy to kinetic energy so that, at some intermediate position, x, its total energy, E, is given by

$$E = E_k + E_p$$
$$= \tfrac{1}{2}mv_x^2 + Mg(h - x)$$

where v_x is the velocity attained at the point x. When it reaches the floor all its energy is kinetic energy and the potential energy is zero. Thus we see that the potential energy describes the position of the mass, that is, its environment.

The total energy, E, of an electron, to include the effect of its environment, must thus be expressed as

$$E = E_k + E_p \tag{2·7}$$

From Eqn (2·5), rearranging

$$k = \frac{2\pi}{h} \sqrt{(2mE_k)}$$

and substituting in Eqn (2·7) we have

$$k = \frac{2\pi}{h} \sqrt{[2m(E - E_p)]} \tag{2·8}$$

Thus we see that the inclusion of potential energy will alter the magnitude of the wave vector **k** when E_p is other than zero.

2·5 Wave equation

All these ideas are brought together in a general equation describing the wave-function, ψ, which is known as Schrödinger's wave equation.

In Eqn (2·3) we saw that the space variation of the wave-function could be expressed as

$$\psi = \exp(jkx) \tag{2·9}$$

Differentiating yields

$$\frac{d\psi}{dx} = jk \exp(jkx)$$

and differentiating again we have

$$\frac{d^2\psi}{dx^2} = -k^2 \exp(jkx) = -k^2\psi$$

Substituting for k^2 from Eqn (2·8)

$$\frac{d^2\psi}{dx^2} + \frac{8\pi^2 m}{h^2}(E-E_p)\psi = 0 \tag{2·10}$$

and this is the wave equation for a wave moving in the x direction. Its solution will yield information about the position of the electron in an environment described by the potential energy, E_p, which itself may be a function of position, x. Equation (2·10) has been 'derived' from the equation for a 'plane' wave (Eqn 2·9) but it does not follow that all wave-shapes will satisfy the same equation. In reality, Eqn (2·10) cannot be derived: it was put forward by Schrödinger as an *assumption*, to be tested by its ability to predict the behaviour of real electrons. A simple example involving solution of the equation for specific conditions is given at the end of the chapter. Other examples are included in the problems. In three dimensions, described by the cartesian coordinates x, y, and z, the wave equation becomes a partial differential equation.

$$\frac{\partial^2\psi}{\partial x^2} + \frac{\partial^2\psi}{\partial y^2} + \frac{\partial^2\psi}{\partial z^2} + \frac{8\pi^2 m}{h^2}[E-E_p(x, y, z)]\psi = 0 \tag{2·11}$$

The solution for ψ will be a function of x, y, and z as also is the potential energy, E_p, and the equation is the time-independent Schrödinger equation. Its solution gives the space-dependent part, ψ, of the wave function, Ψ, and is to be interpreted that the value of $|\psi|^2$ at any point is a measure of the probability of observing an electron at that point. The whole solution can therefore be regarded as a shimmering distribution of charge density of a certain shape. This nebulous solution is in contrast to the sharp solutions of classical mechanics, the definite predictions of the latter being replaced by probabilities.

*2·6 Electron in a box

By way of example of solution of the Schrödinger equation for a given electron environment we consider the case in which it is confined to a specific region of space. In three dimensions this would mean confining the electron in a 'box' but we will treat

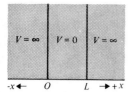

Fig. 2·5 Potential energy defining a 'box'

the problem as one-dimensional with the electron confined in a length, L. We will describe this situation by making the potential energy of the electron zero within the region L and infinity outside it, as illustrated in Fig. 2·5. Since the electron can never acquire infinite energy it is not able to exist where $E_p = \infty$. (In the usual notation employed in wave mechanics the symbol used for potential energy is V and we shall use this here.)

Mathematically we have $V = \infty$ for $x < 0$ and $x > L$, and $V = 0$ for $0 < x < L$. First we consider the solution of the wave equation for the regions outside L. If the electron is to be confined in the region $V = 0$ we require that the probability of finding it outside be zero. Thus our solution to the wave equation must be such that $\psi = 0$ for $x \leqslant 0$ and for $x \geqslant L$. Now the equation is, in one dimension,

$$\frac{d^2\psi}{dx^2} + \frac{8\pi^2 m}{h^2}(E - V)\psi = 0$$

Clearly if V is infinity the term in V must dominate all the others and we can therefore straightaway write

$$-V\psi = 0$$

which, since $V = \infty$, can only be true if $\psi = 0$. Thus the condition of $V = \infty$ outside the length, L, automatically means that there is no probability of the electron being in the regions $x < 0$ and $x > L$, in accordance with our original presumption. It remains to solve the equation for the region $x = 0$ to $x = L$, in which $V = 0$, and the Schrödinger equation is therefore

$$\frac{d^2\psi}{dx^2} + \frac{8\pi^2 m}{h^2}E\psi = 0 \tag{2·12}$$

Since the potential energy of the electron is zero, the energy E in this equation is just the kinetic energy, E_k, which is given by Eqn (2·5). Rewriting that equation we have

$$k^2 = \frac{8\pi^2 m E}{h^2}$$

and substituting in Eqn (2·12)

$$\frac{d^2\psi}{dx^2} = -k^2\psi \tag{2·13}$$

This is a standard form of differential equation the solution to which is described in standard mathematical textbooks and has the form

$$\psi = A\exp(jkx) + B\exp(-jkx) \tag{2·14}$$

This may be tested by differentiation. To evaluate the constants A and B we must consider the boundary conditions of our particular problem. Since we require $\psi = 0$ for $x = 0$ we have, by substituting for ψ and x in Eqn (2·14),

$$0 = A \exp(0) + B \exp(-0)$$

$$= A + B$$

Thus $A = -B$ and Eqn (2·14) may be rewritten as

$$\psi = A[\exp(jkx) - \exp(-jkx)]$$

But $\sin\theta = \dfrac{e^{j\theta} - e^{-j\theta}}{2j}$

so that

$$\psi = 2jA \sin kx \tag{2·15}$$

Now we also require $\psi = 0$ at $x = L$, so that

$$0 = 2jA \sin kL$$

which can only be true if $kL = n\pi$, that is

$$k = \frac{n\pi}{L} \tag{2·16}$$

where n is an integer; therefore

$$\psi = 2jA \sin \frac{n\pi x}{L} \tag{2·17}$$

The question now arises of finding a value for the constant, A, and this is done by what is called *normalization* of the solution. Since $|\psi|^2$ is to be interpreted as a probability and the maximum value of a probability is 1 by very definition of the term, we must arrange that the total sum of $|\psi|^2$ over all space is unity. Expressed mathematically this means that the integral of $|\psi|^2$ over all space equals 1, that is

$$\int_0^\infty |\psi|^2 \, d\tau = 1 \tag{2·18}$$

where $d\tau$ is an element of volume.

In the present problem, 'all space' is confined simply to the length L, so that Eqn (2·18) becomes

$$\int_0^L |\psi|^2 \, dx = 1$$

that is

$$\int_0^L 4A^2 \sin^2\left(\frac{n\pi x}{L}\right) dx = 1$$

Thus,

$$4A^2 \int_0^L \tfrac{1}{2}\left[1-\cos\left(\frac{2n\pi x}{L}\right)\right]dx = 1$$

$$4A^2\left[\frac{x}{2}-\frac{L}{4n\pi}\,\sin\left(\frac{2n\pi x}{L}\right)\right]_0^L = 1$$

Therefore,

$$\frac{4A^2 L}{2} = 1 \quad \text{or} \quad A = \left(\frac{1}{2L}\right)^{1/2}$$

Substituting in Eqn (2·17) gives

$$\psi = j\left(\frac{2}{L}\right)^{1/2}\sin\left(\frac{n\pi x}{L}\right) \tag{2·19}$$

which is the full solution to the equation. From this we may plot $|\psi|^2$, the probability distribution or charge cloud of the electron, as shown in Fig. 2·6(b) for a series of values for the integer, n. Since, from Eqns (2·5) and (2·16)

$$E = \frac{h^2 k^2}{8\pi^2 m} = \frac{h^2}{8\pi^2 m}\frac{n^2\pi^2}{L^2} = \frac{h^2 n^2}{8mL^2} \tag{2·20}$$

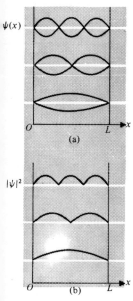

$\psi(x)$

(a)

$|\psi|^2$

(b)

Fig. 2·6 Solutions for an 'electron in a box': (a) the wave function, and (b) the probability as a function of position for different quantum numbers

the energy of the electron increases with n^2. Thus when $n = 2$ the electron has four times the kinetic energy of the electron for which $n = 1$, and so on, and the electron can take only these energy values, i.e., its energy is quantized. It will be seen from the diagram that the effective wavelength of the electron for the case $n = 1$ is such that $L = d/2$, that is, $\lambda = 2L$. For $n = 2$ then $\lambda = L$ and for $n = 3$ we have $\lambda = \tfrac{2}{3}L$ and so on. This is in agreement with our picture of the electron wavelength decreasing as its kinetic energy increases (Eqn. 2·8).

The solution in Eqn (2·19), illustrated in Fig 2·6(a), is exactly like the case of waves on a stretched string—Meldé's famous experiment. The situation of the electron is entirely analagous to the case of a disturbance travelling along a stretched string. When the electron wave reaches the boundary at $x = L$ it is totally reflected and travels back to $x = 0$ where it is again reflected. The result is a standing electron wave with a probability distribution which is stationary in space. The analogy with a stretched string will prove very useful in Chapter 3.

Figure 2·6 shows clearly why the electron energy is quantized: there must always be an integral number of wavelengths in the length $2L$, so the wavelength is constrained to a particular set of values, and with it the energy. It is this feature which makes the results of wave mechanics quite different from those of classical mechanics.

The number n, which determines both the number of half wavelengths making up the length L and the energy, is called a *quantum number.* In subsequent chapters we shall see that in practice there is more than one such number for each electron, and that these quantum numbers have great significance in the explanation of the properties of atoms.

Problems

2·1 Explain why it is necessary to use the *intensity* of a matter wave in order to provide a physical interpretation of the mathematical expression

$$\psi = A \exp[j(\omega t + kx)]$$

2·2 Calculate the magnitude of the wave vector of an electron moving with a velocity of 10^6 m/s.

2·3 If an electron having total energy of 10^{-21} J travels at a constant distance of 100 Å from a proton, what will be its velocity and its value of wave vector?

2·4 If an electron is in a region of constant potential V, show that one solution to the Schrödinger equation

$$\frac{d^2\psi}{dx^2} + \frac{8\pi^2 m}{h^2}(E-V)\psi = 0$$

is $\psi = Ae^{jkx}$, where A is a constant, if

$$k^2 = \frac{8\pi^2 m}{h^2}(E-V)$$

2·5 Show also that if the value of ψ at any point x is equal to the value of ψ at a point $(x+a)$, where a is a constant, then $k = 2n\pi/a$, where n is an integer.

2·6 Why is the function $\psi = $ constant not a permissible solution to problem 2·4? (Hint: consider normalization.)

2·7 Assuming that at very large distances from the origin of axes a wave-function may be written in the form

$$\psi = \frac{c}{r^n}$$

where C is a constant and r is the distance from the origin, show that the wave is only a permissible solution if $n > 2$. (Hint: Consider normalization.)

3 The simplest atom—hydrogen

3·1 Introduction

Now that we have formed a clear picture of how an electron behaves in a rather simple environment we are able to approach the more complicated problem of an electron bound to a single positively charged 'particle'. Such a combination forms a simple atom—indeed, an atom of hydrogen is exactly like this. In it a single electron moves in the vicinity of a particle called a *proton* which carries a positive charge equal in magnitude to that of the electron. Its mass, however, is 1,840 times the mass of the electron so that it behaves much more like a particle than does the electron and we can regard it as a fixed point charge around which travels the electron wave. The shape of this wave could be calculated by solving Schrödinger's equation (Eqn 2·11) with the appropriate expression for E_p. This is mathematically rather difficult so, for the present, we shall make use of the simple analogy noted at the end of Chapter 2 and treat the electron wave as if it were a real wave on, say, a length of string.

Now the electron wave clearly must be three-dimensional but in order to simplify the calculation let us see how far it is possible to go whilst limiting consideration to one dimension only. Later we shall discuss how the results thus obtained are modified by introducing the other two dimensions.

3·2 One-dimensional atomic model

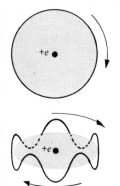

Fig. 3·1 A one-dimensional electron wave orbiting a proton

The simplest model that we can take is illustrated in Fig. 3·1. The electron wave travels around a circular path whose centre is at the atomic nucleus, the proton. Using the vibrating-string analogy of Chapter 2 we can imagine a string stretched around in a circle and set in vibration. Although the path is curved the model is essentially one-dimensional since the amplitude varies only as we move along one direction, that is, the circumference of the circle.

Note that, unlike the case of a stretched string, the waves on it need not be standing waves since there are no ends at which reflections can occur. The wave thus runs around its circular path and in the corresponding electron wave this implies that the mass and charge are being carried around with it. Returning to the string analogy, and guided again by the earlier examples, we may now ask whether the wavelength is in any way restricted.

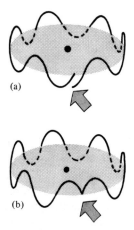

(a)

(b)

Fig. 3·2 Waves with
discontinuous amplitude
or slope are not
permissible

The answer is yes, as can be seen from Fig. 3·2(a) where the wavelength has been arbitrarily chosen with the result that the string does not join up with itself! In electron-wave parlance, we say that the amplitude of the wave must be a smooth and continuous function of position. Figure 3·2(b) shows a case where it is continuous but not smooth and intuitively it seems reasonable that this, too, would not be a stationary state of the system.

The requirement that the string must join up smoothly with itself forces us to the conclusion that there must be an *integral number of wavelengths* in the circumference of the circle. This result leads, not unexpectedly, to quantization of the electron energy as we shall now show.

Assume that there are l wavelengths in the circumference of the circle whose radius is r. Now l can have the values 1, 2, 3, . . . , ∞ (as we shall see later, the value zero is not allowed). This quantization condition may be stated mathematically thus:

$$l\lambda = 2\pi r \tag{3·1}$$

From the wavelength we may obtain the electron momentum, p, using de Broglie's relationship, and combining with Eqn (3·1), we find

$$pr = \frac{lh}{2\pi} \tag{3·2}$$

The product of the linear momentum, p, with the radius, r, is the *moment* of momentum of the electron about the nucleus and is termed the *angular momentum*. We see that it can only have values which are integral multiples of $h/2\pi$, that is, it is quantized in units of $h/2\pi$.

From the momentum, p, it is easy to calculate the kinetic energy, E_k, of the electron as before:

$$E_k = \frac{p^2}{2m} = \frac{l^2 h^2}{8\pi^2 r^2 m} \tag{3·3}$$

Unlike the examples in Chapter 2, the electron also has some potential energy since it moves in the potential distribution of the nuclear charge. This must be added to E_k to obtain the total energy. Now the potential due to the nuclear charge at a distance r is just

$$V = \frac{e}{4\pi\varepsilon_0 r}$$

where ε_0 is the permittivity of free space.

So the potential energy, E_p, of the electron is given by

$$E_p = -eV = -\frac{e^2}{4\pi\varepsilon_0 r} \tag{3·4}$$

In this equation we have used the convention that the electric potential is zero at infinity so that the potential energy of the electron is always negative.

The total energy, E, is obtained by adding Eqn (3·3) and Eqn (3·4) thus

$$E = \frac{l^2 h^2}{8\pi^2 r^2 m} - \frac{e^2}{4\pi\varepsilon_0 r} \tag{3·5}$$

From Eqn (3·5) it does not appear that the energy is quantized. However, we have so far neglected to enquire what is the value of r. We note that as r varies so does the energy but that there is a finite value of r for which the energy is least.

Now it is a universal law of nature that the energy of any system will always try to minimize itself in a given set of circumstances. For instance, a rolling ball will always come to rest at the lowest point of the hill—a position of minimum potential energy. Likewise, a ball which is rolling freely around inside a spherical saucer follows a circular path such that the *sum* of its kinetic energy, E_k and its potential energy, E_p, is a minimum (Fig. 3·3). The radius, R, of its path is thus determined by the condition

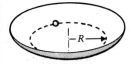

Fig. 3·3 Mechanical model to illustrate energy minimization

$$\frac{dE_k}{dR} + \frac{dE_p}{dR} = 0$$

This condition is exactly equivalent to the statement that the centrifugal force and the inward gravitational force just balance one another, that is, the net radial force on the ball is zero. This can be seen if we remember that energy (or work done) is equal to the integral of force with respect to distance R, that is, $F = dE/dR$. If the total force is to be zero, then $dE/dR = dE_k/dR + dE_p/dR$ must also be zero.

Now we may apply the same principle to the electron and minimize its energy [Eqn (3·5)] with respect to the radius, r:

$$\frac{dE}{dr} = -\frac{l^2 h^2}{4\pi^2 r^3 m} + \frac{e^2}{4\pi\varepsilon_0 r^2} = 0$$

From this equation the equilibrium value of r may be determined,

$$r = \frac{\varepsilon_0 l^2 h^2}{\pi e^2 m} \tag{3·6}$$

By substituting this value of r into Eqn (3·5) the energy under equilibrium conditions is obtained:

$$E = \frac{-e^2 \pi m e^2}{4\pi\varepsilon_0 l^2 h^2 \varepsilon_0} + \frac{h^2 l^2 \pi^2 m^2 e^4}{8\pi^2 m h^4 \varepsilon_0^2 l^4}$$

which simplifies to

$$E = -\frac{e^4 m}{8h^2 \varepsilon_0^2} \cdot \frac{1}{l^2}$$
(3·7)

In exactly the same way as in the example in Chapter 2 we find that the energy is quantized and the permitted values for it are determined by putting $l = 1, 2, 3, \ldots$, in Eqn (3·7). Note that if l were allowed to be zero, the energy would be equal to minus infinity which is physically impossible. There is a useful diagrammatic way of representing *energy levels*, as the permitted values are called. Figure 3·4 shows how energy can be plotted vertically along an axis, and a horizontal line is placed at each value of energy which the electron is allowed to have.

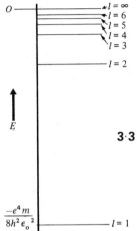

Fig. 3·4 Energy levels for the one-dimensional model of the hydrogen atom

3·3 Atom in two dimensions

Before looking too closely at the results of the previous section we should enquire whether it is to be modified when the restriction to one dimension is removed. Let us now consider the electron wave to be two-dimensional and see how the quantization condition, Eqn (3·1), is to be modified.

We have already seen in Chapter 2 that momentum is a vector quantity, that is, it has direction as well as magnitude, and that its direction in a two-dimensional wave is perpendicular to the wavefront. Since, in general, the wavefront will be curved, so will the path described by the normal to the wavefront. Many such paths may be drawn (Fig. 3·5), all crossing the wavefronts at right-angles and all indicating the direction of the momentum at each point.

The electron wave which encircles the nucleus of an atom may now be imagined in two-dimensional form. Let us concentrate our attention on any one of the many paths followed by the normal to the wavefront. This path must form a closed loop, for,

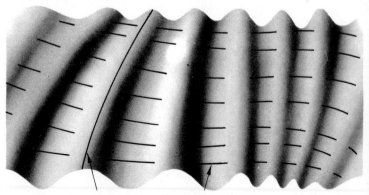

Fig. 3·5 Wavefronts and normals in two dimensions

Wavefront   Normals to wavefronts

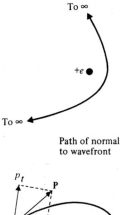

To ∞

+e ●

To ∞

Path of normal
to wavefront

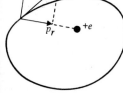

Fig. 3·6 The path
followed by the wave-
front must be closed,
as shown on the right

if it did not, the electron momentum would be carried away from the nucleus towards infinity and the wave would not represent an electron which is localized within the atom [Fig. 3·6(a)]. In Fig. 3·6(b) the electron is confined to the region of space around the nucleus, because its momentum follows a closed path. The illustration also shows the momentum, **p**, at one point, resolved into two components, a radial component, p_r, and a component, p_t, tangential to a circle, centre e.

The wavelength measured along this path must vary from point to point, but since the path is closed there must be an integral number of wavelengths along it, just as in the earlier, one-dimensional calculation. Since the wavelength varies we may express this mathematically by equating the average wavelength, $\bar{\lambda}$, to an integral sub-multiple of the path length, L, that is

$$n\bar{\lambda} = L \qquad n = 1, 2, 3, \ldots \infty.$$

This, then, is our new quantization condition. To simplify the mathematics we may express L in the form $L = 2\pi\bar{r}$, where $\bar{r}$ is now an *average* radius, defined by this equation. Thus

$$n\bar{\lambda} = 2\pi\bar{r} \tag{3·8}$$

is the condition which replaces Eqn (3·1).

If now the kinetic energy and potential energy are expressed in terms of the average values $\bar{r}$ and $\bar{\lambda}$ the total energy E is given by an expression exactly similar to Eqn (3·5).

$$E = \frac{n^2 h^2}{8\pi^2 \bar{r}^2 m} - \frac{e^2}{4\pi\varepsilon_0 \bar{r}}$$

This may be minimized with respect to r exactly as before and the equilibrium energy obtained, now in terms of n instead of l:

$$E = -\frac{e^4 m}{8h^2 \varepsilon_0^2} \cdot \frac{1}{n^2} \tag{3·9}$$

As we shall shortly see, the introduction of a third dimension does not affect this result which therefore gives the energy of the electron in a hydrogen atom. The energy level diagram is exactly as in Fig. 3·4, but with l replaced by n.

The quantum number, l, is not, however, redundant. If we consider now a circular path through the wave motion, it, too, must contain an integral number of wavelengths. Indeed, this must be true of any closed path, whatever its shape, for the wave must always join up smoothly with itself on completion of a circuit. Thus we can retain the quantum number, l, which defines the number of wavelengths in any circular path[†] and

† It is convenient to choose a circular path to define l as it simplifies discussion of the magnetic properties of the atom later on.

which in general is different from n, the number of wavelengths along a different path. The quantum number, l, is related to the *tangential* component of the momentum, p_t, so that we obtain a result similar to that in Eqn (3·2):

$$p_t r = \frac{lh}{2\pi} \qquad (3·10)$$

where r is the radius of the chosen circle. Equation (3·10) then shows that the angular momentum, $p_t r$, is independent of r and is quantized in units of $h/2\pi$. This equation then is an important auxiliary equation which gives further information about the shape of the wave-function. It shows that there is more than one wave-function with the same energy, for, if the quantum number, n, is fixed, different values of l are possible. But the range of permitted values of l is now restricted. We can see this by noting that the component, p_t, of the momentum must always be smaller than the magnitude, p, of the total momentum. This is another way of saying that the wavelength measured along the wavefront normal is the *minimum* value. Thus the path along which n is defined contains the maximum number of wavelengths, and so l *cannot be greater than* n. Moreover, l cannot equal n because in that case the radial component of momentum would be zero and the wavelength measured along the radial direction would be infinite. An infinite radial wavelength implies constant amplitude along the radial direction, *right out to infinity*. Such a state of affairs cannot be if the atom is to have a finite size—in any case, it does not make sense, since the probability of finding the electron, integrated over all space, becomes infinite!

We must conclude, then, that l must always have a value smaller than n. Note that l may take the value zero since this describes a wave with no nodes (that is, constant amplitude) around a circular path and there is nothing to prevent this. It just means that it is equally probable that the electron will be found at any point around the circumference of the circle. Observe also that when l equals zero the electron has no angular momentum—all the momentum is carried in the radial direction. This is not inconsistent with a closed path for the momentum since both the wave and the momentum flow simultaneously outward and inward along the same path, setting up a standing wave pattern.

We may summarize these considerations by stating again the permitted values for n and l, and simultaneously we shall name these quantum numbers:

Principal quantum number, $n = 1, 2, 3, \ldots, \infty$

Angular momentum quantum number, $l = 0, 1, 2, \ldots (n-1)$

Thus if the electron is a wave-function whose principal quantum number is 3 then l may have the values 0, 1, 2 only. These three wave-functions all have the same energy,

$$E = -\frac{e^4 m}{8h^2 \varepsilon_0^{\ 2}} \times \frac{1}{9}$$

When $n = \infty$ the electron has zero energy and is at an infinite distance from the nucleus—it may be considered to have been removed from the atom, and the amount of energy needed to remove it, starting from its lowest energy level, is just $e^4 m/8h^2\varepsilon_0^2$, equal to 13·6 electron volts. A smaller energy would be required if the electron were initially in a higher level.

The importance of these quantum numbers will become clearer in the next chapter where atoms containing many electrons will be studied. For the reader who is already familiar with the Bohr theory of the atom, in which the electron is treated as a particle, Section 3·8 shows how the wave theory is related to it and points out its shortcomings.

3·4 Three-dimensional atom—magnetic quantum numbers

It might be expected that the introduction of the third dimension would involve quantization of the component of momentum perpendicular to the plane of the two-dimensional wave. This approach, however, is mathematically difficult and does not give us as much physical insight as the method we shall follow.

The electron wave possesses angular momentum about the nucleus which implies that the electron mass is circulating with the wave. The electron charge is intimately associated with its mass, so it, too, must be circulating as if there were an electric

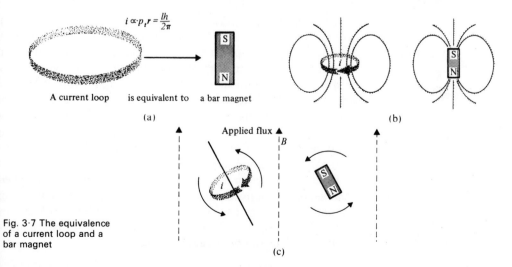

$$i \propto p_t r = \frac{lh}{2\pi}$$

A current loop   is equivalent to   a bar magnet

(a)

(b)

Applied flux B

Fig. 3·7 The equivalence of a current loop and a bar magnet

(c)

Path followed by tip of spindle

Anticlockwise couple due
to weight of gyroscope and
reaction of support.

Fig. 3·8 The motion of a
gyroscope under a
gravitational couple is
called 'precession'

current flowing in a loop around the nucleus (Fig. 3·7). Indeed, the equivalent current is directly proportional to the angular momentum.

Such a current loop behaves exactly like a small permanent magnet. It has the same distribution of magnetic flux around it [Fig. 3·7(b)] and behaves in a similar way when in a magnetic flux applied from an external source. Thus the two-dimensional electron wave experiences a couple on it when a magnetic flux, **B**, is applied as shown [Fig. 3·7(c)]. The couple acts in such a way as to try to turn the plane of the wave perpendicular to the flux lines. The wave is not free to so turn, however, because its circulating mass makes it behave like a gyroscope. A gyroscope under the action of the couple due to gravity (Fig. 3·8) does not fall over but *precesses* about the direction of the gravitational field. Likewise the electron wave precesses about the direction of an applied magnetic flux (Fig. 3·9).

This kind of precession must virtually always be present since it is rare that the magnetic flux in an atom is exactly zero although it may often be very weak. If there is no flux from an external source then that due to the nucleus, which is also a small magnet, is sufficient to cause precession of the electron wave.

Now this precessing wave forms a three-dimensional wave-function and, just as in the two-dimensional case, any closed path through it must contain an integral number of wavelengths. Indeed, any path may be chosen in order to define a third quantum number but the most convenient and useful one is a circle around the flux axis. The angular momentum about this axis is therefore quantized exactly—as is the total angular momentum—in units of $h/2\pi$. We shall use the symbol m_l for the new quantum number so that the angular momentum about the flux direction is $m_l h/2\pi$.

Since the total angular momentum is $lh/2\pi$, the momentum about the flux axis must be a component of it—in fact we can represent angular momentum in vectorial form as we did linear momentum. The vector representing the total angular momentum is drawn along the axis of rotation, its length proportional to the magnitude of the angular momentum. Figure 3·10 shows this and we see that this vector precesses about the flux axis with a *constant component* along that axis. This component therefore must be equal to $m_l h/2\pi$. The other component rotates about that axis and is not quantized in any simple way.

From this diagram it is easy to see that the *magnetic quantum number*, m_l, has a maximum value equal to l, when l points along **B**, and the inset in Fig. 3·10 shows how the vector $m_l h/2\pi$ can point in the opposite direction to **B**. This is described mathematically by giving m_l a negative value.

Cone swept out by axis
normal to plane of 2D wave

Fig. 3·9 The electron
wave (represented here
as a current loop)
precesses around the
field direction

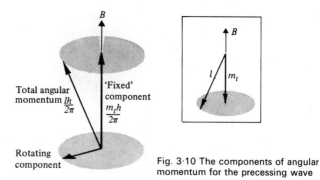

Fig. 3·10 The components of angular momentum for the precessing wave

The range of values permitted for m_l thus extends from $-l$ through zero to $+l$. If, for example, $l = 3$, then m_l can have one of the values -3, -2, -1, 0, 1, 2, 3—seven possibilities in all.

3·5 Spin of the electron

It is of interest to note that each dimension in the above problem is associated with a quantum number and that this was also true in the cases discussed in Chapter 2. It is indeed generally true that three quantum numbers are necessary to describe a wave-function.

So far, however, we have not considered the possibility of motion within the electron itself. The wavelike model of the electron does not include this possibility but it has been found necessary to assume that the electron is spinning internally about an axis in order to explain many aspects of the magnetic properties of atoms. Pictorially we may represent the spinning electron as fuzzy distribution of mass rotating about its own centre of gravity rather as a planet spins on its axis while orbiting the sun. It is difficult to couple this image with that of the wave motion around the nucleus and the reader is advised not to try to do so! However, a spinning wave packet is not too difficult to imagine although more difficult to draw.

As might be expected, the angular momentum of this spinning motion is quantized, and in units of $h/2\pi$, too. To distinguish it from the *orbital* angular momentum, $lh/2\pi$, we call it the *spin* angular momentum and its magnitude is $m_s h/2\pi$, where m_s is the spin quantum number. Unlike l and m_l, though, m_s does not have integral values but can only take one of the two values $+\frac{1}{2}$ and $-\frac{1}{2}$. Thus the motion can be either clockwise or anticlockwise about the axis, the spin angular momentum being equal to $\pm\frac{1}{2}h/2\pi$. Since the vector representing this points along the axis of rotation, the two states are often referred to as 'spin up' and 'spin down' (Fig. 3·11). An explanation of the reason for the non-integral values of m_s is beyond the scope of this book.

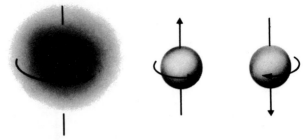

Pictorial spinning electron   Equivalent motion of a 'particle electron'

Fig. 3·11 Pictorial spinning electron (left) and equivalent motion of a 'particle electron'

3·6   Electron clouds in the hydrogen atom

Since the shape of the wave-function is defined by the quantum numbers n, l, and m_l, so too is the probability density distribution for the electron. Thus we may imagine an 'electron cloud' which represents the charge distribution in a hydrogen atom just as described in Chapter 2. Such a cloud can be illustrated as in Fig. 3·12, where some examples are given which represent different states of the hydrogen atom. The quantum numbers

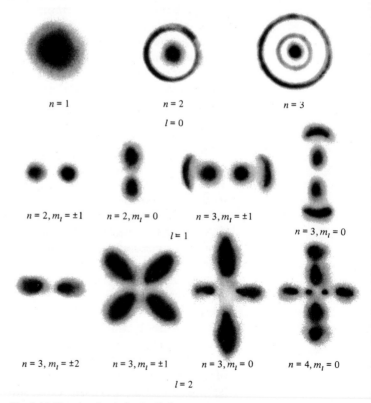

Fig. 3·12 Electron clouds for the hydrogen atom

corresponding to each state are given below the illustrations. Each cloud is shown in cross-section and the density of the image is proportional to the charge density. These clouds all have rotational symmetry about the vertical axis, which coincides with the axis along which m_l is quantized. Note that since the electron is not a standing but a running wave the probability density does not vary along a circle described around that axis since it is the time-average probability.

The $l = 0$ states are all seen to be perfectly spherical in shape, while the others are quite asymmetrical. These shapes will be referred to again when we consider atomic bonding in Chapter 4.

3·7 Energy levels and atomic spectra

One of the most direct checks on the accuracy of the expression for electron energy is to measure the energy released when the electron 'jumps' from an energy level to a lower one. In Fig. 3·13 such a transition between energy levels is represented by a vertical arrow: its length is proportional to the energy released.

Since no matter is expelled from the atom during the transition the only form which the emitted energy can take is an electromagnetic wave—a photon. The photon energy must be exactly equal to the difference in energy between the two electron energy levels concerned so it can be calculated from Eqn (3·9). (The possibility exists that two photons might be emitted but the chance of this is so small that it is negligible.)

If the electron makes a transition between the levels with quantum numbers n_1 and n_2 the photon frequency may thus be found from the equation

$$E_{\text{photon}} = hf = -\frac{e^4 m}{8h^2 \, \varepsilon_0^2 n_1^2} + \frac{e^4 m}{8h^2 \, \varepsilon_0 n_2^2}$$

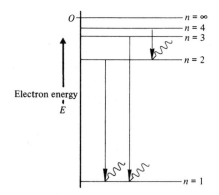

Fig. 3·13 Transitions between energy levels involve the emission of photons

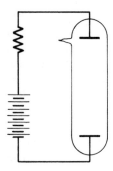

Fig. 3·14 A discharge
tube and its power
supply

that is, $\quad f = \dfrac{e^4 m}{8h^3 \, \varepsilon_0^2}\left(\dfrac{1}{n_2^2} - \dfrac{1}{n_1^2}\right)$ \hfill (3·11)

The frequency may thus have any one of a discrete set of values obtained by substituting different values for n_1 and n_2.

How can we arrange to observe these? The first requirement is a source of hydrogen in which there are many atoms in *excited* states, that is, states other than the one with lowest energy (called the *ground* state). Fortunately this is easily achieved by setting up an electric discharge in gaseous hydrogen. In a discharge tube (of which the neon tube frequently used for advertisements is an example) a large potential difference is applied between two electrodes (Fig. 3·14). Once the discharge has been started the gas glows brightly and a large current flows. The current is carried across the gas by ions (that is, atoms which have lost or gained an electron and which are hence charged). The potential gradient across the tube accelerates these ions to high velocities whereupon they collide with neutral atoms and excite the latter into states of high energy. As these excited atoms return to the ground state they emit photons as described above.

The emitted light contains several wavelengths and the corresponding frequencies are given by Eqn (3·10) above. The light from a discharge tube is thus normally coloured.

The wavelengths present may be separated and measured using a spectrometer. In its simplest form, this has a glass prism to split the light into its components but more often a diffraction grating is used. Figure 3·15 shows how the beam transmitted by a narrow slit is diffracted onto a photographic film. From a knowledge of the spacing of the grating lines and Bragg's law the wavelengths may be measured. In Fig. 3·15 we also show a photographic record of the spectrum of hydrogen obtained in this way.

It is found that Eqn (3·10) accurately predicts the measured wavelengths.†

If the discharge tube is placed in a large magnetic field during the measurement the quantum states with different values of m_l and m_s have different energies. The highest energy occurs when the total angular momentum vector points along the field direction and the lowest when it points in the opposite direction. These shifts in the energy levels shown in Fig. 3·13 may be observed experimentally and confirm the predictions of our theory. It was through such measurements of spectral lines that the theory was originally built up, and the existence of the spin of the electron was first established.

---

† Even more accurate agreement is obtained when corrections are made for the finite mass of the nucleus and for relativistic effects.

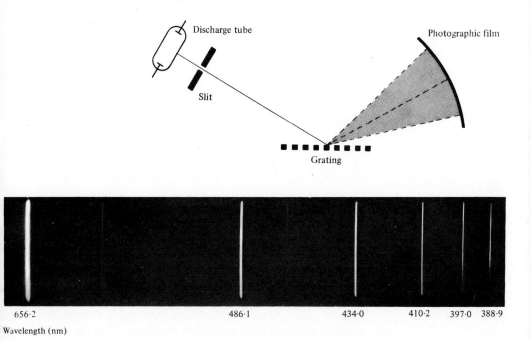

Fig. 3.15 Apparatus for recording the emission spectrum of a gas, and a spectrum of hydrogen obtained in this way

*3·8 Bohr theory of the atom

Following the suggestion that the atom may be like a planetary system as outlined in Chapter 1, Bohr calculated the energy levels of such a model. He assumed that the electron was a particle which orbited the nucleus along a circular path (Fig. 3·16) at a distance r from it.

The attractive force between the electron and the proton is $e^2/4\pi\varepsilon_0 r^2$. The inward acceleration is equal to v^2/r, where v is the velocity, so that Newton's third law of motion gives

$$\frac{e^2}{4\pi\varepsilon_0 r^2} = \frac{mv^2}{r} \tag{3.12}$$

As mentioned in Chapter 1, the accelerating electron is expected to radiate energy. In order to get round this, Bohr had to postulate that particular orbits were stable non-radiating states. In order to obtain agreement with the spectral lines emitted he proposed an arbitrary quantization condition: the angular momentum should be an integral multiple of $h/2\pi$. There is no justification for this condition without introducing the wave nature of the electron; this is the fundamental weakness of the theory from which its other failures stem.

The quantization condition is thus

$$mvr = \frac{nh}{2\pi}, \qquad (n = 1, \ 2, \ 3, \ \ldots) \tag{3.13}$$

We wish to obtain an expression for the total electron energy, which is just

$$E = \tfrac{1}{2}mv^2 - \frac{e^2}{4\pi \, \varepsilon_0 r}$$

Using Eqn (3.12) and Eqn (3.13) both v and r can be eliminated from this expression, with the result

$$E = -\frac{me^4}{8\varepsilon_0^2 h^2 n^2} \tag{3.14}$$

which is identical to Eqn (3.9).

Soon after Bohr's calculation Sommerfeld extended the theory to include the possibility of elliptical orbits. To do this he introduced a second quantum number, k, which is analogous to the angular momentum quantum number l, and retained the quantum number n to give the total momentum of the electron. Unfortunately the model leads to incorrect values for the angular momentum, as follows. When $k = n$, all the momentum is angular momentum, that is, the orbit is a circle, as in Bohr's original theory. As we saw earlier, this result is not allowed by the wave theory but it cannot be excluded in the present theory. On the other hand, the case $k = 0$ means that there is no angular momentum, that is, the 'orbit' must be a straight line passing through the nucleus which on physical grounds should be impossible. However, we know from the wave theory that the angular momentum may indeed be zero, as this does not imply collision of the electron with the nucleus.

In spite of the shortcomings which we have mentioned the Bohr–Sommerfeld theory of the atom enjoyed considerable success. Corrections were made to account for the finite mass of the nucleus (both the electron and the nucleus must rotate about their common centre of mass) and for relativistic effects, where the electron velocity is large. These changes brought the theory into very close agreement indeed with the energy levels deduced from spectroscopic measurements.

However, it could not deal with the criticisms mentioned above nor with the fact that many of the expected spectral lines either do not appear or are extremely weak. In all these respects the wave theory has proved entirely satisfactory.

Problems

3·1 A ball of mass 1 g moves in a circular path on the inside surface of an inverted cone, in a manner similar to the example in Fig. 3·3. If the apical angle of the cone is 90°, find an expression for the energy levels of the ball, assuming its wavelength to be given by De Broglie's relation. Hence show that the quantization of its energy may be neglected for practical purposes.

3·2 Calculate the average diameter of the wave-function of the hydrogen atom in its ground state. Compare it with the distance between the atoms in solid hydrogen, which may be calculated from its density, $76\cdot3$ $kg\,m^{-3}$.

3·3 Calculate the energies of the first three levels of the hydrogen atom. What are the frequencies of radiation emitted in transitions between those levels? Find the first transition for which the radiation is visible, if the shortest visible wavelength is 400 nm.

3·4 How many different values may the quantum number m_l have for an electron in each of the states $n = 3$, $l = 2$; $n = 4$, $l = 2$; $n = 4$, $l = 3$?

3·5 Deduce a formula for the energies of the helium ion, He^+, in which a single electron moves around a nucleus whose charge is $+2e$. (*Hint:* although the nuclear charge is doubled, the electronic charge is not, so that it is not correct to replace e by $2e$ in Eqn 3·9.)

3·6 Discuss the consequences if Planck's constant were to have the value 10^{-3} joule-sec.

3·7 Draw a rough plot of the amplitude of the electron wave versus distance from the nucleus for the state $n = 3$, $l = 0$ illustrated in Fig. 3·12. Do the same for the state $n = 3$, $l = 2$, $m_l = 0$, along the vertical axis. How many maxima are there in each case?

3·8 Explain why in this figure, the state $n = 2$, $l = 1$, $m_l = 0$ has two maxima around a circumference, while the fact that $l = 1$ indicates that there is only *one* wavelength in the same distance.

3·9 What do you expect might happen if an electron with kinetic energy greater than 13·6 eV were to collide with a hydrogen atom?

3·10 If the space around the nucleus of a hydrogen atom were filled with a dielectric medium with relative permittivity ε_r, calculate the new expression for the energy levels. Hence find the energy of the level $n = 1$ when $\varepsilon_r = 11\cdot7$.

(The relevance of this question to semiconductor theory will be discussed in Chapter 13.)

4 Atoms with many electrons—the Periodic Table

4·1 Introduction—the nuclear atom

In the early years of this century the structure of atoms had to be laboriously deduced from a varied collection of facts and experimental results. It was easily discovered that atoms of most elements contained several electrons since they acquired charge in multiples of $+e$ when electrons were knocked out of the atom by collision. Because an atom is normally neutral, therefore, it must also contain a number of positive charges to balance the negative charges on the electrons. From the experiment by Rutherford mentioned in Chapter 1 it was also found possible to deduce the actual charge on the atomic nucleus by analysing the paths of the deflected α-particles. In this way Rutherford showed that the number of positive charges of magnitude e on the nucleus just equalled the atomic number—the number assigned to an element when placed with the other elements in sequence in the Periodic Table. As the reader who has studied more advanced chemistry will know, this sequence is nearly identical to that obtained by placing the elements in order of increasing atomic weight and has the merit that the relationships between elements of similar chemical behaviour are clearly displayed. This is a topic which we shall cover later in this chapter.

Now if the nucleus of an atom contains Z positive charges each equal in magnitude to the electronic charge it follows that the neutral atom must contain Z electrons. By analogy with the case of hydrogen we therefore anticipate that these electrons are in motion around the nucleus, bound to it by the mutual attraction of opposite charges. It remains only to assign appropriate quantum numbers to each electron and we shall then have a model of the atom with which it will prove possible to explain many of its properties, including chemical combination. This is the aim of this and the next few chapters.

But before proceeding, note one point that is as yet unexplained. We have already identified the proton as a fundamental particle and we expect an atomic nucleus to contain Z such protons. However, the mass of an atom is much greater than the mass of Z protons so there must be some other constituent of the nucleus. In any case, it is unreasonable to expect a group of protons, all having the same charge, to form a stable arrangement without some assistance. The extra ingredient has been identified as an electrically neutral particle of nearly the same mass as the proton, called the *neutron*. Several neutrons are found in each nucleus: the number may vary without changing the chemical properties of the atom. The only quantity which changes is the atomic weight, and this explains the existence of different isotopes of an element as mentioned in Chapter 1.

Since the proton and neutron have nearly identical masses and the electron masses may be neglected in comparison, the atomic weight A must be nearly equal to the total number of protons and neutrons. So the number of neutrons is just the integer nearest to $A-Z$. The force which binds the uncharged neutrons to the protons is a new kind of force, called *nuclear force*, which is very strong compared to the electrical repulsion between the protons.

Since the structure of the nucleus has little or no bearing on the chemical and physical properties of an element, we shall not study it further.

4·2   Pauli exclusion principle

In assigning quantum numbers to the Z electrons in an atom the first consideration must be that the atom should have the minimum possible energy. For if one electron could make a transition to a lower energy level it would do so, emitting radiation on the way. At first sight this implies that the electrons must all be in the lowest energy level, each having the same quantum numbers, that is, $n = 1$, $l = 0$, $m_l = 0$, and $m_s = +\frac{1}{2}$. But this is not so and, indeed, the truth is almost exactly the opposite. It is found that each electron has its own set of quantum numbers which is different from the set belonging to every other electron in the atom. This result has become enshrined in a universal principle named after Wolfgang Pauli, who first deduced it. In its simplest form, Pauli's exclusion principle states that:

no more than one electron in a given atom can have a given set of the four quantum numbers.

To emphasize this important principle we state it again in a different way: no two electrons in an atom may have all four quantum numbers the same.

The origin of this principle may be partly understood from the following argument. The three quantum numbers n, l, and m_l completely determine the shape of the wave function and hence also the charge distribution. If two electrons have the same values for these quantum numbers their charge distributions are thus directly superimposed on one another, or in other words the electrons are 'in the same place'. Now this is a very unfavourable situation, since the two electrons must repel one another very strongly. It occurs in practice only if the electron spins are opposed: their quantum numbers, m_s, are then different. It seems that two electrons with counter-rotating spins do not repel one another very strongly. On the other hand, if their spins are parallel the repulsion is so strong that they cannot have the same position (charge distribution) and therefore they have a different set of values for n, l, and m_l.

4·3 Electron states in multi-electron atoms

We may now use the exclusion principle to assign quantum numbers for the first few atoms in the table of elements and we shall find that we are led quite naturally to the construction of a table which reflects the chemical similarities and differences between the elements.

The element with atomic number, Z, of two is helium. It contains two electrons, and at least one of the quantum numbers of the second electron must differ from those of the first. On the other hand both electrons must have the lowest possible energy. Both these requirements are met if the electrons have the following quantum numbers.

$$\text{1st electron:} \quad n = 1 \quad l = 0 \quad m_l = 0 \quad m_s = +\tfrac{1}{2}$$
$$\text{2nd electron:} \quad n = 1 \quad l = 0 \quad m_l = 0 \quad m_s = -\tfrac{1}{2}$$

This places both electrons in states with the same wave function, but with opposite spins.

Atomic number $Z = 3$ corresponds to lithium. Two of the three electrons may have the sets of quantum numbers given above for helium. The third electron must have

$$n = 2 \quad l = 0 \quad m_l = 0 \quad m_s = +\tfrac{1}{2}$$

Here we note that the lowest energy for this electron is the level $n = 2$, which is higher than the energy level with $n = 1$. This should mean that this electron can be removed more readily from the atom than either of the electrons in the helium atom, since less energy is needed to take the electron away to an infinite distance. This is, indeed, the case, for while helium is a noble gas, lithium is metallic, and we know that metals readily emit electrons when heated in a vacuum while helium certainly does not.

In assigning the value $l = 0$ to this third electron we have made use of another rule—*in a multi-electron atom the levels with lowest l values fill up first*. This may be stated in another way: the energy of an electron increases with l as well as with n.

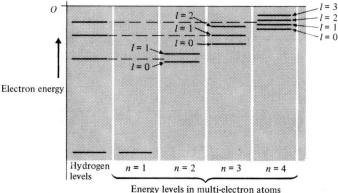

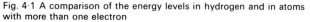

Fig. 4·1 A comparison of the energy levels in hydrogen and in atoms with more than one electron

Remember that this was not so in the hydrogen atom; the expression for the energy [Eqn (3·9)] contained only the quantum number, n. Now, however, the electrical repulsion between the electrons alters the energy of each level as indicated in Fig. 4·1, splitting what was a single level in the hydrogen atom into a series of levels each with a different value for the angular momentum quantum number, l.

4·4   Notation for quantum states

Before going further we shall introduce a shorthand notation for the quantum numbers and their values. The principal quantum number, n, defines a series of levels (often called 'shells' because each level corresponds to a different mean radius, $\bar{r}$, of the wave function), each shell corresponding to one value of n. Each of these shells is assigned a letter according to the scheme:

$$n: \quad 1 \quad 2 \quad 3 \quad 4 \ldots$$
$$\text{Letter}: \quad K \quad L \quad M \quad N \ldots$$

Similarly, the values of l are characterized by another series of letters, which, like the above, derive from the early days of spectroscopy:

$$l: \quad 0 \quad 1 \quad 2 \quad 3 \ldots$$
$$\text{Letter}: \quad s \quad p \quad d \quad f \ldots$$

All energy levels belonging to given values of n and l are said to form a subshell and these are labelled by the number corresponding to the value of n and the appropriate letter for the value of l. Thus the subshell with $n = 3$, $l = 2$ is denoted by $3d$, that with $n = 2$, $l = 0$ is denoted by $2s$.

We now continue to build up the electronic structure of the elements assuming that the levels are filled sequentially with rising values of n and l. Table 4·1 shows the number of electrons in each subshell for each of the first 18 elements.

Table 4·1   Electrons in each sub-shell for the first 18 elements

Atomic Weight (A)	Atomic Number (Z)	Element	K	L		M		
			$1s$	$2s$	$2p$	$3s$	$3p$	$3d$
1·008	1	H	1					
4·003	2	He	2					
6·94	3	Li	2	1				
9·01	4	Be	2	2				
10·81	5	B	2	2	1			
12·01	6	C	2	2	2			
14·01	7	N	2	2	3			
16·00	8	O	2	2	4			
19·00	9	F	2	2	5			
20·18	10	Ne	2	2	6			
22·99	11	Na	2	2	6	1		
24·31	12	Mg	2	2	6	2		
26·98	13	Al	2	2	6	2	1	
28·09	14	Si	2	2	6	2	2	
30·97	15	P	2	2	6	2	3	
32·06	16	S	2	2	6	2	4	
35·43	17	Cl	2	2	6	2	5	
39·95	18	A	2	2	6	2	6	

Note that there is a maximum number of electrons that can be put into each subshell, and by looking at the details of quantum numbers we can see why this is.

For instance in the K (or $1s$) shell, we have $n = 1$, $l = 0$, $m_l = 0$. There are two possibilities for m_s, these are $+\frac{1}{2}$ and $-\frac{1}{2}$. Hence only two electrons are permitted in the K shell.

In the L shell, where $n = 2$, the possible values for l, m_l, and m_s are $l = 0, 1$; $m_l = -1, 0, +1$; $m_s = \pm\frac{1}{2}$. So the possible combinations of these label the various 'states' into which electrons may go:

$$\left.\begin{array}{lll} l = 0, & m_l = 0, & m_s = +\frac{1}{2} \\ l = 0, & m_l = 0, & m_s = -\frac{1}{2} \end{array}\right\} 2 \text{ states in the } 2s \text{ subshell}$$

$$\left.\begin{array}{lll} l = 1, & m_l = 0, & m_s = \pm\frac{1}{2} \\ l = 1, & m_l = 1, & m_s = \pm\frac{1}{2} \\ l = 1, & m_l = -1, & m_s = \pm\frac{1}{2} \end{array}\right\} 6 \text{ states}$$

making six states in the $2p$ subsheil, and a grand total of eight in the L shell.

It is easy to derive the corresponding numbers for subsequent shells from the rules:

for each value of n there are n possible values of l;

for each value of l there are $(2l+1)$ possible values of m_l;
for each value of m_l there are two possible values of m_s.

Thus the total numbers of electrons which can be accommodated in each succeeding shell are found to be

$$K : 2 \qquad L : 8 \qquad M : 18 \qquad N : 32$$

4·5 Periodic Table

Certain interesting features arise in Table 4·1. The elements with atomic numbers 2, 10, and 18 are the rare gases which are not merely stable but are exceedingly inert chemically. Thus we may equate inert characteristics with completely filled shells of electrons. It will be noted that the outer shell contains two electrons in helium and eight in both neon and argon. This is the first indication, as we move down the list of elements, that similar arrangements of outer electrons appear periodically and that this periodic variation is reflected in the chemical properties. In fact the periodicity of chemical properties was noted before the electronic structure of the elements was known and as long ago as 1870 the chemist Mendeléev devised a way of tabulating the elements which demonstrated it very effectively. This is now known as the Periodic Table of the Elements.

To construct the Periodic Table we take the elements listed in Table 4·1 and place them in order in horizontal rows so that the outer electron structure changes stepwise as we proceed along the row. We begin a new row whenever a p shell has just been filled with electrons. Ignoring for the moment elements one and two this means that each row finishes with a rare gas, for example, neon with its full L shell or argon with full $2s$ and $2p$ shells. In this way elements with identical numbers of electrons in their outermost shells appear directly beneath one another. Thus lithium, sodium, and potassium, each with one electron outside full shells (usually called *closed* shells) appear in the first column. More remarkably, these elements all display very similar chemical behaviour—they are all very reactive, are metals with a valency of unity, and they form similar compounds with, for example, fluorine or chlorine. We may amplify this last point by noting the properties of the chlorides: they all form transparent, insulating crystals, which are readily cleaved to form regular and similar shapes. They all dissolve, to a greater or lesser extent, in water and they all have fairly high melting points (about 700°C).

We note corresponding similarities between the elements in the second column: beryllium, magnesium, and calcium are all light metals with a valency of two; they form very stable oxides which have even higher melting points than the alkali halides mentioned above and tend to be reactive although not to the degree shown by the alkali metals.

It would be possible to fill a book by listing all the properties which are shared by elements in the same column but by now the reader should be able to recognize that this unity of chemical behaviour is common to all the columns (or *Groups*, as they are called) of the Periodic Table. Moreover, it is reasonable to associate similar chemical behaviour with a similarity in the occupancy of the outermost electron shells. The importance of these outer electrons is brought home further when we observe that the principal valency of each Group is related to the number of such electrons. Thus in Groups I, II, and III the valencies are identical to the numbers of outer electrons, while in Groups V, VI, and VII the valencies are equal to the *number of electrons which would have to be added to complete the outermost shell*. We shall return to this feature later, for this is our first glimpse of the role of electrons in chemical combination.

4·6 Transition elements

The reader will observe that we have so far been very careful to limit discussion to the first three rows of the Periodic Table plus potassium and calcium. The reason is that while the periodicity of properties continues beyond this point it is not exemplified in such a simple fashion.

Let us study the filling of energy levels in the elements of the fourth row. These are shown in Table 4·2 (note the new notation), and immediately an anomaly is apparent. The rule concerning the order of filling the various levels has been broken! Instead of the outermost electrons of potassium (K) and calcium (Ca) entering the 3*d* shell, they go into the 4*s* shell. Only when the 4*s* shell is full do electrons begin to enter the 3*d* shell—there is one 3*d* electron in scandium, two in titanium, and so on.

Table 4·2   Arrangement of electrons for elements 19 to 29

Z	Element	Electron Configuration†			
19	K	(Filled K and L shells)	$3s^2$	$3p^6$	$3d^0$	$4s^1$
20	Ca	(Filled K and L shells)	$3s^2$	$3p^6$	$3d^0$	$4s^2$
21	Sc	(Filled K and L shells)	$3s^2$	$3p^6$	$3d^1$	$4s^2$
22	Ti	(Filled K and L shells)	$3s^2$	$3p^6$	$3d^2$	$4s^2$
23	V	(Filled K and L shells)	$3s^2$	$3p^6$	$3d^3$	$4s^2$
24	Cr	(Filled K and L shells)	$3s^2$	$3p^6$	$3d^5$	$4s^1$
25	Mn	(Filled K and L shells)	$3s^2$	$3p^6$	$3d^5$	$4s^2$
26	Fe	(Filled K and L shells)	$3s^2$	$3p^6$	$3d^6$	$4s^2$
27	Co	(Filled K and L shells)	$3s^2$	$3p^6$	$3d^7$	$4s^2$
28	Ni	(Filled K and L shells)	$3s^2$	$3p^6$	$3d^8$	$4s^2$
29	Cu	(Filled K and L shells)	$3s^2$	$3p^6$	$3d^{10}$	$4s^1$

†In the notation used here the superscript indicates the number of electrons which occupy the sub-shell.

The reason for this oddity is that the 4s levels have *lower* energy than the 3d levels so that they fill first in keeping with the minimum energy principle. This is illustrated in the energy diagram in Fig. 4·1, where the positions of the energy levels in the hydrogen atom are shown for comparison. The shift in the relative energies in a multi-electron atom is another example of the way the interactions between electrons modify the wave-functions and their energies. The more electrons there are competing for space around the nucleus, the more important these interactions become. So, while the interactions have been small up to this point, from now on we shall find more and more examples of interchanged energy levels.

Returning to the filling of the 3d shell in elements 21 to 29 we note the irregularities at 24 chromium (Cr) and 29 copper (Cu), each of which contains one 4s electron instead of two. This is due to the fact that the exactly half-filled 3d shell and the filled 3d shell are particularly stable configurations (that is, they have lower energy) compared to the neighbouring occupancies of four and nine electrons respectively.

How do we assign these elements to their correct Group in Periodic Table?

Table 4·3   Arrangement of electrons for elements 29 to 36

Z	Element	Electron Configuration†		
29	Cu	(Filled K, L, M shells)	$4s^1$	
30	Zn	(Filled K, L, M shells)	$4s^2$	
31	Ga	(Filled K, L, M shells)	$4s^2$	$4p^1$
32	Ge	(Filled K, L, M shells)	$4s^2$	$4p^2$
33	As	(Filled K, L, M shells)	$4s^2$	$4p^3$
34	Se	(Filled K, L, M shells)	$4s^2$	$4p^4$
35	Br	(Filled K, L, M shells)	$4s^2$	$4p^5$
36	Kr	(Filled K, L, M shells)	$4s^2$	$4p^6$

†In the notation used here the superscript indicates the number of electrons which occupy the sub-shell.

In spite of the complications, it is clear that the first two elements, potassium (K) and calcium (Ca), fall respectively into Groups I and II both on grounds of chemical similarity and because of the similarity of the 'core' of electrons remaining after removal of the valence electrons. Assignment of subsequent elements to their groups is easier if we first discuss 29 copper and succeeding elements up to number 36 where we arrive at krypton, another inert gas, with a stable octet of outer electrons. The intervening elements, 29 to 35, all have complete inner shells and unfilled outer shells; they thus fall naturally into Groups I to VII in sequence. However, copper does not fit quite so naturally with the very reactive lithium (Li) and sodium (Na) because removal of an electron leaves not a stable core but the configuration $3d^9\ 4s^1$ of nickel (Ni). We therefore divide Group I (and, for similar reasons, Group II) into subgroups labelled A and B, putting Cu and Zn into the latter.

Let us now return to the elements 21 scandium (Sc) to 28 nickel (Ni). This set is called the first series of *transition elements*, which, because of the presence of the 4s electrons, all have similar properties (they are all metals). However, the first five fit quite naturally into Groups II–VII, though again we find it appropriate to divide the groups into subgroups, putting the transition metals into the A subgroups. But for the elements Fe, Co, and Ni there are no precedents in the table, and these we assign to a new Group (VIII).

Having dealt with the first series of transition elements, we are not surprised to find another series in the fifth row of the table, and also in the sixth. The latter, however, is more complicated owing to the filling of two inner subshells (4f and 5d) before the transition is complete and the 6p shell begins to fill. Here, the series of elements 57 to 71, in which the 4f shell is being filled, have almost identical chemical properties—as a Group they are called the *rare earth metals*. The reason for their chemical similarity is that these elements differ only in the number of electrons in a shell which is far removed from the outermost electrons while it is the latter which determine chemical behaviour. Their almost indistinguishable qualities place them all into one pigeon-hole in the Periodic Table, in Group IIIA.

The subsequent filling of the 5d shell beyond its solitary occupancy in 71 lutecium (Lu) marks the continuation of the Periodic Table, for the 5d electrons, like the 3d electrons in the first transition series, are of chemical importance, so that the properties of these elements differ.

Another group of elements similar to the rare earths is found in the seventh periodic row, although the majority of these do not occur in nature because their nuclei are unstable—they have, however, been manufactured artificially. The elements in this last series in Group IIIA are often referred to as the *actinide elements* (they all behave like actinium, the first of the series) just as the rare earths are sometimes called *lanthanide elements*.

4·7 Valency and chemical combination

It is necessary to clarify some points concerning valency, particularly in the transition elements. We have already hinted that valency is determined by the ease with which an atom may become ionized, that is, either lose or gain electrons in its outer shells. Thus magnesium (Mg) and calcium (Ca) are divalent because the two s electrons, known as the *valence* electrons, are readily removed from the neutral atoms. Removal of a third electron would involve destruction of the completed and very stable octet of electrons in the next lowest shell. The stability of closed s and p shells is thus displayed in all atoms, not just in the rare gases which happen to possess closed shells in the electrically neutral state.

By contrast, a transition element deprived of its outermost s electrons (for example, the two $4s$ electrons in iron) is not nearly as stable. It may easily lose yet more electrons, and can consequently have more than one valency. Thus we find ions (atoms with more or less than the normal number of electrons) such as Fe^{3+} and Fe^{2+} (the superscript gives the resultant charge on the ion). The Fe^{2+} ions would be found in a compound such as ferrous oxide (FeO) or ferrous bromide ($FeBr_2$) whilst Fe^{3+} occurs in ferric oxide (Fe_2O_3) and ferric bromide ($FeBr_3$).

In these examples the oxygen and bromine atoms have valencies of two and one respectively because they require the addition of this number of electrons to attain a stable configuration with closed outer shells. Since in this stable configuration the outer shell contains eight electrons the number of electrons to be added is $8-N$, where N is the group number. Thus the principal valency of the elements in Groups IV–VII is given by the so-called $8-N$ rule.

Hydrogen is a unique case since the nearest stable arrangement contains two electrons (a filled K shell), or alternatively no electrons. Hydrogen atoms can either lose or gain electrons so permitting combination with halides to form acids (HF, HCl, HBr, HI) and also with metals, giving hydrides (NaH, ZrH, CaH, and so on).

The noble gases are assigned to Group zero: their valency is nominally zero and they were thought until quite recently to be completely inert. It is now known that they form a few simple compounds and that they can also combine with themselves to form solids but only at very low temperatures.

It is of interest to note at this point that although the individual wave-functions of electrons in a subshell have quite complex shapes (cf. Fig. 3·12), in a completed subshell the total charge density is quite simply spherically symmetric. The electron cloud of a filled subshell is thus like that of one of the s states in Fig. 3·12 and a noble gas atom may therefore be pictured roughly as a small sphere.

Problems

4·1 Discuss the chemical similarities which lead us to place the elements of subgroups IIA and IIB together in Group II. Discuss also the differences between these subgroups.

4·2 Plot the melting points and densities against atomic number for the first three rows of the Periodic Table. Notice how the changes reflect the periodicity represented by the Table. (Data in Appendix III)

4·3 Look up the densities of the elements in Group IIA and explain why they depend on atomic number.

4·4 What differences would there be in the Periodic Table if (a) the $5p$ levels had lower energy than the $4d$ levels; (b) the $5d$ levels had lower energy than the $4f$ levels?

4·5 In which elements do the nuclei contain the following numbers of neutrons?

  10   14   22

Suggest how the existence of several isotopes might explain why chlorine apparently has the non-integral atomic weight 35·4.

4·6 Write down the quantum numbers of each electron in the M shell and hence show that it may accommodate only 18 electrons.

4·7 Why do the oxides of the Group II metals have higher melting points than the alkali halides? Why are the Group II metals less reactive than those in Group I?

4·8 Without consulting Table 4·1, write down the electronic configuration of the elements with atomic numbers 4, 7, 10, 15. By counting the valence electrons, decide to which Group each element belongs.

4·9 Copper has a single valence electron in the $4s$ shell and belongs to Group I. What is the electronic structure of the Cu^+ ion? Why does copper not behave chemically in the same way as sodium or potassium?

4·10 The outermost shells of all the inert gases are filled s and p subshells. Silicon and germanium have four electrons outside closed s and p shells, and so have titanium and zirconium. Why, then, are the chemical characteristics of all these elements not identical?

5 Molecules and bonding

5·1 Molecular bond

When a gas of neutral atoms condenses to form a solid the atoms are held rigidly together by mutual attraction. The pull between them is much stronger than a mere gravitational force and we say that atoms form *bonds* between one another. In a solid each atom forms a bond with all of its neighbours but in a diatomic molecule the single bond may be studied in isolation. We therefore commence our discussion of the mechanism of bond formation with a study of the molecular bond.

5·2 Formation of a molecule

Just as the simplest atom to study is hydrogen so the simplest molecule is formed by the combination of two hydrogen atoms—the diatomic hydrogen molecule.

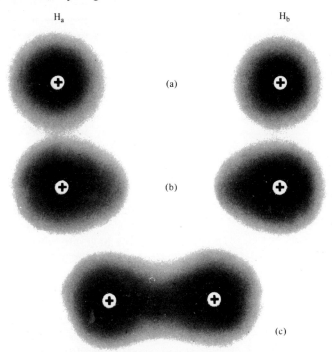

Fig. 5·1 (a) Isolated atoms; a weak attractive force is present; (b) at closer approach the electron clouds are distorted by the attractive forces, the two electron spins are antiparallel; (c) in equilibrium the electrons share the same wave function

This combination of two protons and two electrons is not so easy to treat mathematically as is the single atom. However, a qualitative picture may easily be obtained by imagining the atoms H_a and H_b to approach one another as depicted in Fig. 5·1. As their separation decreases each electron feels two new competing forces: an attraction towards the other nucleus and a repulsion from the other electron. In practice the attraction is the stronger, for the repulsion is small if the electrons have opposite spins. This can be seen from Pauli's exclusion principle which allows electrons to have the same wave-functions, that is, occupy the same region of space, only if their spins are anti-parallel.

The atoms are thus pulled together until each electron cloud surrounds both nuclei at which stage the two electrons share the same wave-function [Fig. 5·1(c)]. Closer approach is impossible because the repulsion between the two nuclei would be too great. At the equilibrium separation, however, there is an exact balance between the attraction of the electrons and the repulsion of the nuclei. This is shown in Fig 5·2 where each force is plotted against the interatomic distance. As the atoms approach both attraction and repulsion forces build up, the latter more slowly at first. On close approach the repulsive force increases rapidly until it equals the attraction and the molecule is in equilibrium. In hydrogen this occurs with a separation d_0 of 0.74 Å. Thus it is possible to calculate the equilibrium separation if the dependence of the forces on separation is known.

Note that bonding results in a marked overlap of the two electron clouds. This is a universal characteristic and it means that a pair of atoms form a bond only when they approach close enough for the valence electron clouds to overlap one another.

When two heavier atoms combine there is one small difference from the case discussed.

Consider the approach of two sodium atoms each of which we may picture as a core of negative charge (the completed K and L shells) outside of which moves the valence electron cloud (Fig. 5·3). We have already seen that the closed shells of the core are spherical in shape so that they are represented as spheres in the illustration. The bonding force arises exactly as in the hydrogen molecule but the repulsive force is now augmented at short separations by the reluctance of the closed inner shells to overlap. They could overlap only if the electron spins were opposed (see the discussion of Pauli's principle, Chapter 4) but since each closed shell contains electrons with spins already opposed, overlap is not allowed. Hence the equilibrium separation is determined largely by the radius of the inner core and the molecule is very like a pair of nearly rigid spheres

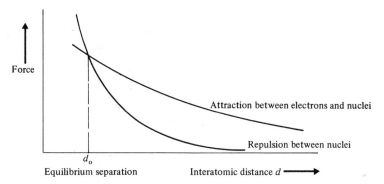

Fig. 5·2 Forces between two atoms plotted against interatomic distance

glued together. This simple model is of great utility in considering the way atoms pack together in crystals (cf. Chapter 6).

5·3 Bonds in solids

We have seen how the overlapping of the valence electron clouds of two atoms can lead to an overall attraction and hence to the formation of a molecule. The formation of a solid body may be described in similar terms but with attraction between each atom and all of its neighbours simultaneously instead of with just one or two other atoms. The question of how many neighbours surround each atom and in what arrangement is left to the next chapter. Here we shall look primarily at the bond formed between one pair of atoms in a solid but we must remember that both atoms form more than just this one bond.

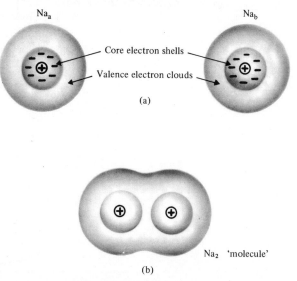

Fig. 5·3 The formation of a 'molecule' of Na. This structure is not stable in practice; however, the addition of further atoms produces a stable crystal of sodium—see later

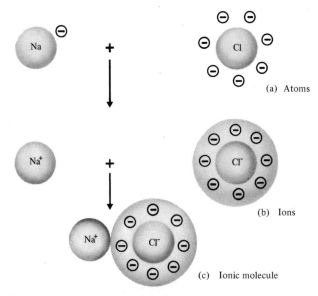

(a) Atoms

(b) Ions

(c) Ionic molecule

Fig. 5·4 Schematic representation of the formation of an ionic molecule' of sodium chloride

Bonds between atoms may vary in nature according to the electronic structure of the atoms concerned. One would not expect the closed-shell structure of the inert gases to behave in the same way as the alkali metals, each of which can so easily lose its outer electron. Indeed, we know from experiment that while these metals readily form compounds with other elements, the inert gases do not.

There are actually five classes into which bonds are conventionally divided, although the boundaries between these categories are not too well defined, as we shall see.

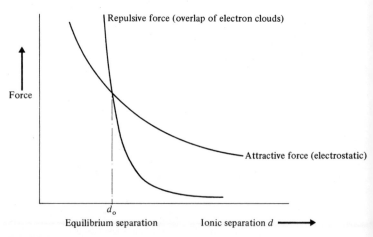

Fig. 5·5 Dependence of inter-ionic forces on ionic separation

5·4   Ionic bonding

As remarked above, the valence electron is very easily removed from an alkali metal leaving behind a very stable structure resembling an inert gas but with an extra positive nuclear charge.

On the other hand an element from Group VII (fluorine, chlorine, bromine, or iodine—the halides) is but one electron short of an inert gas structure. Since the inert gas structure is so stable we might expect that a halide atom would readily accept an extra electron and might even be reluctant to lose it given suitable conditions.

Taking these two tendencies together we can understand what happens when, for example, sodium and chlorine atoms are brought together in equal numbers. It is easy for the valence electron of each sodium atom to be transferred to a chlorine atom, making both more stable. But now there must be an electrostatic attraction between the ions so formed for each sodium ion carries a positive charge (Fig. 5·4) and each chlorine ion a negative one. The attraction brings them closer and closer together until the inner electron clouds begin to overlap. At this point a strong repulsion force is manifested exactly as for the case of two sodium atoms described earlier and the two forces just balance one another. The way in which these forces depend upon the separation, d, is shown in Fig. 5·5, and we see that it is very similar to the case of the hydrogen molecule, Fig. 5·2.

Naturally there is also an electrostatic repulsion between ions of the same charge so that in solid sodium chloride (NaCl) we do not expect to find like ions side by side but rather alternating as shown in Fig. 5·6. In this way the attraction of unlike ions overcomes the repulsion of like ions and a stable structure is formed. We shall meet this and other kinds of atomic packing in Chapter 6.

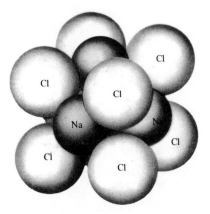

Fig. 5·6 Part of the crystal of NaCl modelled as a close packed arrangement of spheres

In Fig. 5·6 we represent the ions by spheres for we have already seen (Fig. 5·3) that this is quite a realistic way of picturing the inner core of electrons.

It is to be noted that while the structure shown in Fig. 5·6 is stable, a single pair of ions do not form a stable molecule—the molecule of sodium chloride does not normally exist by itself.

Magnesium oxide (MgO) has a similar structure to that of sodium chloride but in this case two electrons are transferred from each magnesium atom to an oxygen atom, again giving each ion a stable group of eight outer electrons (Fig. 5·7). Since these ions are doubly ionized and hence carry two electronic charges it is understandable that the interatomic cohesion should be much stronger than in sodium chloride. This accounts for the much higher melting point of magnesia (2,800°C compared with 800°C for sodium chloride). The relationships between melting point and bond strength will be discussed later in this chapter but it suffices here to note that the stronger a bond the higher is the temperature needed to break it.

Yet other examples of ionic bonding occur in the compounds cupric oxide, chromous oxide, and molybdenum fluoride (CuO, CrO_2, MoF_2) showing that the metallic element need not be from Groups I or II but that any metal may readily become ionized by losing its valence electrons.

5·5   Covalent bonding

Elements from the central groups of the Periodic Table—notably Group IV—are not readily ionized. The energy required to remove all the valence electrons is too large for

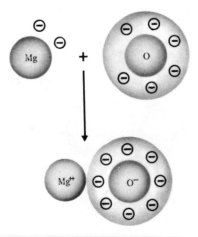

Fig. 5·7 Representation of the formation of an ionic molecule of MgO

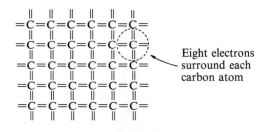

Eight electrons surround each carbon atom

Fig. 5·8 By sharing electrons (shown by dashes) with its neighbours, a carbon atom forms four bonds, and a repetitive structure can be created

ionic bonding to be possible. It is still possible for each atom to complete its outer shell, however, by sharing electrons with its neighbours.

We saw earlier how the hydrogen molecule is formed by sharing electrons in this way but we now show that this bond is more universal and can be present in solids.

We may take carbon as an example. It has a filled K shell and four electrons in the L shell, that is, $1s^2 2s^2 2p^2$. Four more electrons are required to fill the L shell and these may be acquired by sharing an electron with each of four neighbours when carbon is in its solid form. One way in which this could be done is shown in Fig. 5·8 although this arrangement does not occur in practice because of the directional nature of the bonds.

To understand this we must look at the shape of the electron clouds in a carbon atom. The four electrons in the L shell interact rather strongly with one another, modifying the shapes of the s and p wave-functions until they are virtually identical. The electrostatic repulsion is so strong that each electron's 'cloud' concentrates itself away from those of the other three. As illustrated in Fig. 5·9, this means that each cloud is sausage-like and points away from the nucleus, the four arranging themselves with the largest possible angle, 109·5°, between each pair. It is easy to see that they point towards the corners of an imaginary tetrahedron.

Because of the strong repulsion this arrangement is difficult to distort and the carbon atoms join up as shown in Fig. 5·9 with the electron clouds of neighbouring atoms pointing towards one another. In this way each carbon atom is surrounded by eight electrons which is a stable arrangement. The structure so formed is very strong and rigid—it is the structure of diamond, the crystalline form of carbon.*

*The term crystal (from the Greek *kristallos*—clear ice) until recently referred to a solid which can be cleaved to form a regular geometrical shape, or which occurs naturally in such a shape. The modern definition of a crystal will be given. in Chapter 6.

Fig. 5·9 (a) the lobe-shaped valence electron clouds in the carbon atom; (b) the carbon atom in its crystalline surroundings has a similarly shaped electron cloud

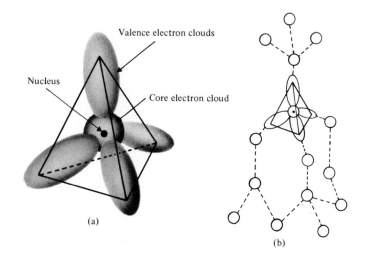

(a)

(b)

Just as in the case of sodium chloride, no molecule can be distinguished here but the solid is rather like one huge molecule (a so-called macromolecule) since it is a never-ending structure. One can always add more carbon atoms.

The other Group IV elements, silicon and germanium, also crystallize in the same structure and compounds of an element of Group III with one of Group V form the related zincblende structure which we shall meet in Chapter 6. In these compounds each atom is again tetrahedrally coordinated with four others even though the isolated Group III and Group V atoms have three and five valence electrons respectively.

Here we should note how the valency of covalent elements may be deduced. In forming the covalent bond the atom acquires electrons by sharing until it has a stable outer shell. The valency thus equals the number of electrons lacking from the outer shell; if there are N electrons present in the outer shell of the neutral atom then the valency is $8-N$.

5·6 Metallic bonding

As its name implies, metallic bonding is confined to metals and near-metals many of which are to be found in Groups I, II, and III of the Periodic Table. If we take as an example copper, we see that shells K, L, and M are full, while there is just one $4s$ electron in the N shell. In solid copper the outer electron is readily released from the parent atom and all the valence electrons can move about freely between the copper ions. The positively charged ions are held together by their attraction to the cloud of negative electrons in which they are embedded (Fig. 5·10) rather like ball bearings in a liquid glue.

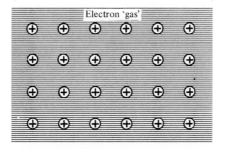

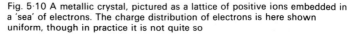

Fig. 5·10 A metallic crystal, pictured as a lattice of positive ions embedded in a 'sea' of electrons. The charge distribution of electrons is here shown uniform, though in practice it is not quite so

In some respects this arrangement is like an ionically bonded solid but instead of the negative 'ions' being like rigid spheres they fill all the space between the positive ions. One might say that a metallic bond is a sort of ionic bond in which the free electron is donated to all the other atoms in the solid.† Thus the bonds are not directional and their most important characteristic is the freedom of the valence electrons to move. In later chapters we shall see how this mobility of the electrons is responsible for the high electrical and thermal conductivities of metals.

Because the ions are bonded to the valence electrons and only indirectly to each other it is possible to form a metallic solid from a mixture of two or more metallic elements, for example, copper and gold. Moreover, it is not necessary for the two constituents to be present in any fixed ratio in order that the solid be stable. The composition may thus be anywhere between that of gold with a small proportion of copper added and of copper containing a small proportion of gold. It is rather as if one metal were soluble in the other and, indeed, we often refer to this type of material as a solid solution. More commonly the term *alloy* is used and in Chapter 10 we shall learn more about alloys, some of which are of great importance because of their great strength and lightness.

5·7 Van der Waals bonding

The inert gases condense to form solids at sufficiently low temperatures although one would not expect them to form bonds of any of the above kinds. Similarly, valency requirements are fulfilled in molecules of the gases methane, carbon dioxide, hydrogen, and so on (CH_4, CO_2, H_2, etc.) and no spare

† Although this is the easiest way to understand the metallic bond it is often useful and legitimate to view it as a kind of covalent bond in which the electrons form a temporary covalent bond between a pair of atoms and then move on to another pair.

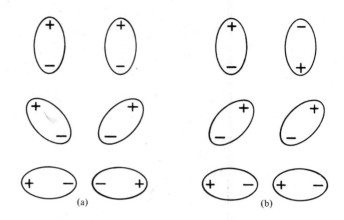

Fig. 5·11 Illustrating the origins of the Van Der Waals force

electrons are available for forming bonds, yet they too can solidify. It is clear that there must be another kind of bond, often called a secondary bond, in which no major modification of the electronic structure of atoms takes place.

We know that an atom with closed electron shells consists essentially of a positive nucleus surrounded by a spherical cloud of negative charge. If the electron clouds of two such atoms were static and undeformable no force would exist between them. Actually, however, the electron clouds result from the motion of the various electrons and although it is not possible to follow their trajectories the electrons must be regarded as in motion around the nucleus. Thus whilst an atom can on average have no electrical dipole moment it has a rapidly fluctuating dipole moment. At any instant the centre of the negative charge distribution does not coincide with the nucleus but rapidly fluctuates about it.

When, therefore, two atoms approach one another the rapidly fluctuating dipole moment of each affects the motion of the electrons in the other atom and a lower energy (that is, an attraction) is produced if the fluctuations occur in sympathy with one another. This can be seen by considering the alternative situations shown in Fig. 5·11, where each atom is shown as a small electric dipole. In Fig. 5·11(a) we see several situations in which little or no attraction is produced, while in Fig. 5·11(b) all cases display a marked attractive force. The fluctuations occur in such a way that the situations like those of Fig. 5·11(b) occur most frequently so that a net attractive force exists.

Clearly this force acts even over large distances for the electron clouds do not need to overlap one another. As a result it makes itself felt even in the gaseous state, and causes deviations from the perfect gas laws. The well-known gas equation suggested by van der Waals incorporates its effects and the force is thus known after him.

We shall come across many substances which are bonded by van der Waals forces: they include plastics, graphite, and the paraffins.

5·8   Hydrogen bond

There is another secondary bond which can be made between atomic groups which have no electrons to spare. It bears a close resemblance to the van der Waals bond which, as we saw, is due to the attraction between dipoles, but occurs only when a hydrogen atom is present.

Consider an ionic molecule formed by an atom of hydrogen and some other atom or group of atoms which we shall call X. We may describe the structure by the symbol X^-—H^+ (The signs $+$ and $-$ indicate the charge on the ions.) This molecule is a small dipole. The positive charge on the hydrogen ion can attract the negative charge on a third ion or group of atoms, which may even be part of another molecule. Thus we have a weak dipole–dipole attraction exactly as in the van der Waals bond. But the H ion is very small since it has lost its valence electron and the third ionic group, Y, may approach very closely, forming a slightly stronger bond than otherwise. We may denote this structure symbolically

$$X^-\text{—}H^+\text{———}Y^-$$

Note that, since the bond with Y is not a true electronic bond, Y cannot approach as closely as X to the hydrogen ion. However, it would be quite possible to have the structure

$$X^-\text{———}H^+\text{—}Y^-$$

in which the ionic bond appears between Y and H while the X—H bond is now weak. In practice the hydrogen atom hops backwards and forwards between the two possible positions (near to X and near to Y) so the bond must be described as a mixture of the two forms. The net attractive force between X and Y is rather weak since it cannot be stronger than its weakest part (the dipolar attraction). Modified forms of this bond occur when hydrogen is present in both of the molecules which form the bond.

The hydrogen bond, as it is called, occurs frequently in organic materials in which hydrogen often plays a major role. We shall meet it in Chapter 11 for it can make an important contribution to the strength of plastics (polymers). As might be expected, it also appears in water and ice and it explains the

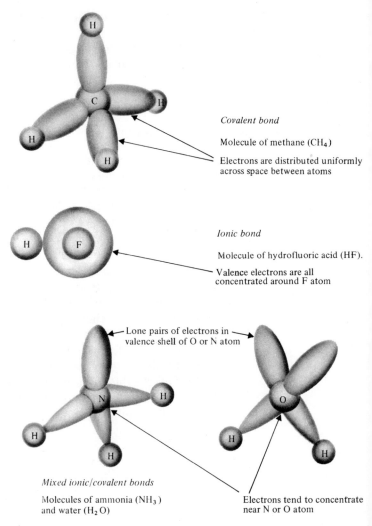

Covalent bond

Molecule of methane (CH_4)

Electrons are distributed uniformly across space between atoms

Ionic bond

Molecule of hydrofluoric acid (HF).

Valence electrons are all concentrated around F atom

Lone pairs of electrons in valence shell of O or N atom

Mixed ionic/covalent bonds

Molecules of ammonia (NH_3) and water (H_2O)

Electrons tend to concentrate near N or O atom

Fig. 5.12 Electron clouds for covalent, ionic, and mixed ionic-covalent bonds

occurrence of such hydrated inorganic salts as nickel sulphate ($NiSO_4 . 6H_2O$).

5·9 Comparison of bonds—mixed bonds

Both the van der Waals bond and the hydrogen bond differ in a quite distinctive way from the three primary bonds, and they give quite different properties to a solid. In the following sections we shall compare the properties of the five bonds, taking some examples to illustrate them.

Although at first sight the three primary bonds seem quite different the bonding in many substances does not fit easily into just one or other of these categories.

Let us consider a series of molecules which contain the same total number of electrons, and which are hence called *isoelectronic*: methane, ammonia, water, and hydrogen fluoride.

In methane (CH_4) the carbon atom can form four covalent bonds, one with each hydrogen atom (Fig. 5·12). In hydrogen fluoride (HF) at the other extreme, the hydrogen donates its electron to the fluorine atom to complete the L shell of the latter, and the atoms are bonded ionically together. In the intervening compounds, ammonia and water (NH_3 and H_2O), neither of these pictures is quite correct in itself but a mixture of the two applies. In such cases a proportion of an electronic charge may be considered to be transferred while the rest of it is shared equally to form a partial covalent bond. In reality this means merely that the electron clouds are distorted roughly as shown in Fig. 5·12 so that there is a predominance of electronic charge density away from the hydrogen atom. This shift of the 'centre of charge' increases as we go through the series from the C—H bond to the F—H bond. The carbon, nitrogen, oxygen, and fluorine atoms are arranged in order of increasing 'electronegativity' which is defined as the degree to which an atom can attract electrons to itself. In general, the electronegativity increases steadily across the Periodic Table from Group I to Group VIII. It also varies within a Group, decreasing with increasing atomic number except in the case of transition elements.

If we are given the electronegative series, we can thus form rules about bonding:

Two atoms of similar electronegativity form either a *metallic* bond or *covalent* bond, according to whether they can release or accept electrons.

When the electronegativities differ the bond is partially ionic, the ionic character increasing with the difference in electronegativity.

Since the covalent bond is directional, while the ionic bond is not, the degree of directionality changes with the bond character. Such changes have a marked influence on the crystal structure as we shall see in Chapter 6.

A further example of mixed ionic/covalent bonding occurs in an important class of substances based upon the dioxide of silicon (SiO_2). This compound can form both a regular, ordered structure—in which form it is called quartz—and an irregular structure called quartz glass or silica. Many of the glasses and various ceramic materials, for example porcelain, are complex structures containing silicon dioxide and other compounds in which regular structures are embedded in an amorphous† matrix. Such materials will be discussed in more detail in Chapter 10.

Just as bonds occur with mixed ionic/covalent natures, so we find that there is a continuous change in bonding character in a series of alloys of metals such as Cu–Ni, Cu–Zn, Cu–Ga, Cu–As, and Cu–Se. In fact, the last few of these tend to form definite compounds (that is, the elements combine in simple ratios), indicating the presence of a bond other than metallic. However, at the same time they retain certain characteristically metallic properties.

5·10   Bond strength

Obviously, because there is a smooth transition between covalent and ionic and metallic bonds there is also a smooth variation in bonding strength. However, if we confine our attention to pure covalent, ionic, hydrogen, and van der Waals bonds, some characteristic distinctions emerge.

The strength of a bond is best measured by the energy required to break it, that is, the amount of heat which must be supplied to vaporize the solid and hence separate the constituent atoms. If we measure the heat of vaporization of one kilogramme mole this must be the energy used in breaking N_o bonds ($N_o =$ Avogadro's number). Figures derived in this way are tabulated in Table 5·1 together with the type of bonding for the solid concerned. As might be expected, the weakest are the van der Waals bond and the hydrogen bond. Next comes the metallic bond, followed by the ionic and covalent bonds whose strengths are nearly comparable.

5·11   Bond strength and melting point

Melting points are also indicative of bond strengths for the following reason. The atoms in a solid are in constant vibration at normal temperatures. In fact the heat which is stored in the material when its temperature is raised is stored in the vibrational energy of the atoms which is proportional to the (absolute) temperature. Thus melting occurs when the thermal vibration becomes so great that the bonds are broken and the atoms become mobile.

By comparing melting points it is therefore possible to illustrate some interesting variations of bond strength. For

---

† *Amorphous*—having no form or structure (from the Greek *morphe*—form). Amorphous solids do not form crystals with regular shapes because there is no regularity in the way in which their atoms are packed together. Note that not all solids which do not form regular geometrical crystals are amorphous.

Table 5·1

Bond Type	Material	Heat of vaporization H (kcal/mole)
van der Waals	He	0·02
	N_2	1·86
	CH_4	2·4
Hydrogen:		
O—H——O	Phenol	4·4
F—H——F	HF	6·8
Metallic	Na	26
	Fe	94
	Zn	28
Ionic	NaCl	153†
	LiI	130†
	NaI	121†
	MgO	242†
Covalent	C (Diamond)	170
	SiC	283
	SiO_2	405
	Si	85

†Not strictly the heat of vaporization, since the solid does not break down into its constituent monatomic gases when vaporized. These figures are thus obtained by adding those for a sequence of several changes of state with the required end result.

instance we have already remarked upon the extra strength conferred on the ionic bond by high valency, and how this results in the high melting point of magnesium oxide compared to sodium chloride. It is also clear from the following list of melting points of Group IV elements how bond strength decreases with increasing atomic number.

Material	Atomic Number	Melting Point (°C)
Carbon (diamond)	6	3,750
Silicon	14	1,421
Germanium	32	937
Tin	50	232

We conclude from these and other data that electrons of higher principal quantum number form weaker bonds.

Bond strength is clearly of importance in determining the strength of materials although the form of the crystal structure has at least as much, if not more, influence on strength. Thus we shall leave a discussion of this to a later chapter.

In some solids two kinds of bonding are separately present: the solid forms of the gases are a case in point. Nitrogen (N_2), carbon dioxide (CO_2), oxygen (O_2), and nitrous oxide (NO_2) all have covalent or mixed ionic–covalent bonds within the molecule.

In solid and in liquid form the molecules are held together by van der Waals forces. This explains how it is that they solidify only at very low temperatures since the van der Waals bond is easily broken by thermal agitation. On the other hand these gases break down into their constituent atoms (that is, dissociate) only at very high temperatures because of the strong intramolecular bonds. Similar behaviour is found in polymers (see Chapter 11) and in many organic materials.

Problems

5·1 Using the data in Table 5·1, calculate the energy in eV required to break a single bond in MgO.

5·2 Calculate the angles between the tetrahedral bonds of carbon.

5·3 The distance between the atoms in the HF molecule is 0·917 Å. Calculate the maximum attractive force between two HF molecules separated by a distance of 10 Å and compare this with the attractive force between the H^+ ion and the F^- ion when they are separated by 10 Å.

5·4 What kind of bonding do you expect in the following materials: GdO, GdTe, SO_2, RbI, FeC, C_6H_6, InAs, AgCl, UH_3, GaSb, CaS, BN, Cu-Fe?

5·5 The attractive force between ions with unlike charges is $e^2/4\pi\varepsilon_0 r^2$ while the repulsive force may be written Ce^2/r^n, where the exponent n is about 10 and C is a constant.

Obtain an expression for the equilibrium distance r_0 between the ions in terms of C and n. Hence deduce an expression for the energy E required to separate the ions to an infinite distance apart, and show that it may be written in the form

$$E = \frac{e^2}{4\pi\varepsilon_0 r_0}\left(1 - \frac{1}{n-1}\right)$$

(*Hint:* energy = integral of force with respect to distance.)

5·6 To remove an electron to infinity from an Na atom, 5·13 eV of energy must be used, while adding an electron from infinity to a Cl atom to form the Cl^- ion liberates 3·8 eV of energy. What is the energy expenditure in transferring an electron from an isolated sodium atom to an isolated chlorine atom?

Using the expression given in problem 5·5, calculate also the energy change when the two isolated atoms are brought together to form an NaCl molecule, given that the ionic separation is 2·7 Å at equilibrium, and that the exponent n in the expression for the energy is equal to 10. Compare your result with the figure given in Table 5·1 for the heat of sublimation. Comment on the discrepancy between the two figures.

5·7 Does the decrease in melting point with increasing atomic number which is displayed by the elements in Group IV appear in all the Groups? (Data in Appendix III.)

6 Crystal structure

6·1 Introduction

It was mentioned in the last chapter that the properties of a solid depend as much upon the arrangement of atoms as on the strength of the bond between them. We met there some examples of such arrangements, diamond and rock salt (NaCl), and also an irregular solid, glass. We know that the regular structures often lead to regular crystalline solid shapes, as in the precious stones, and that amorphous solids do not form crystalline shapes.

Between these extremes lie other categories such as the metals which show regularity in the way their atoms pack together but which do not normally take on crystalline shapes. The concept we use to distinguish the regularity or otherwise of the atomic packing is that of *order*. Thus in a diamond crystal (Fig. 5·9) there is perfect order because each atom is in the same position relative to its neighbours, irrespective of where in the crystal it lies. We say that the crystal has *long range order*.

By contrast, an amorphous solid displays no order—it is completely disordered.

In yet other cases the long range order exists over distances of only 0·1 mm or so. This is still very long compared to the atomic spacing (2Å). So the term 'long range' still applies. We illustrate this kind of structure in two dimensions in Fig. 6·1. Each region of long range order is called a crystallite or grain and the solid is termed polycrystalline. Metals, some ceramics, and many ionic and covalent solids are often found in this form.

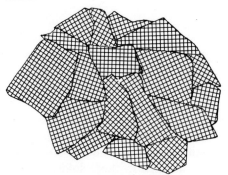

Fig. 6·1 A polycrystalline solid is composed of many grains oriented at random and separated by grain boundaries

30μm

Fig. 6·2 A photomicrograph of an etched polycrystalline specimen of copper showing grain boundaries

The boundary between crystallites is called a grain boundary, and it is at this surface that the regularity 'changes direction'. Grain boundaries can be shown up by etching a metal with a suitable acid—the metal dissolves more readily at the boundary —and a micrograph of an etched polycrystalline specimen of copper is shown in Fig. 6·2. The dimensions of the grains can be anything upwards of about 50Å across.

To distinguish those solids in which the order extends right through we call them either single crystals or monocrystalline solids.

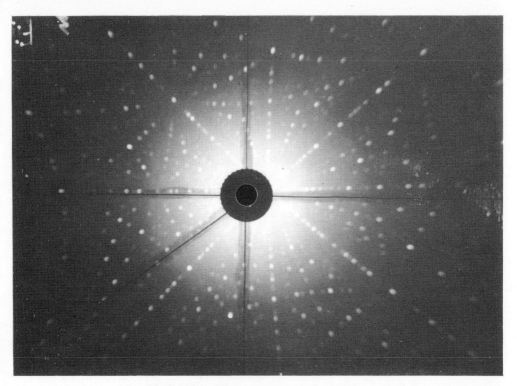

Fig. 6·3 X-ray diffraction pattern obtained photographically from a single crystal of Cr_2O_3. The dark lines and the central disk are not produced by the X-rays.

Yet another situation arises in materials like glass in which order extends over distances embracing a few atoms only. This we call short range order and it is also found in liquids. Indeed, many solids which display short range order are actually super-cooled liquids and are not strictly in equilibrium in this state. Glass will sometimes actually crystallize if given sufficient time (several hundred years) or if it is exposed to shock.

Polymers also display order but in a unique way and we leave discussion of their structure to a later chapter.

6·2 Single crystals

Now that we have met the four kinds of solid—monocrystalline, polycrystalline, glassy, and amorphous—we can go on to study the different forms of crystalline order. Historically, crystallography began in about 1700 with the study of the external form or morphology of crystals but it was only in the first years of this century that the concept of a crystal as a regular array of atoms was established. Then in 1912 the diffraction of X-rays by crystals was discovered. X-rays are like light waves of very short wavelength, about a few ångstroms. Since this is comparable to the interatomic spacing in a crystal the X-ray is readily diffracted and this property can be used for studying the arrangement of atoms and their spacing. The wavelength of the

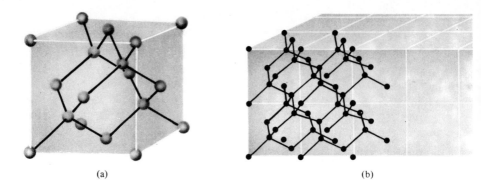

(a)   (b)

Fig. 6·4 (a) The unit cell of the diamond structure; (b) the crystal of diamond is built by stacking many unit cells together

X-rays can be independently measured by recording their diffraction by a ruled grating.

In Fig. 6·3 we show a diffraction photograph taken using the arrangement shown in Fig. 6·15 of a single crystal of chromium oxide (Cr_2O_3). The exposed dots on the photographic plate are at the intersections of the diffracted beams with the plate. The analysis of such a photograph to determine the crystal structure is very complex but the techniques for doing this are now well established. Later in this chapter we shall describe the experimental method and prove Bragg's law. In the meantime we consider only the results showing atomic arrangements which have been deduced from much patient analysis of X-ray photographs. We shall emphasize the way in which the choice of crystal structure follows from consideration of the bonding requirements.

First we must define more carefully what is meant by crystalline order. In a crystal it is possible to choose a small group of atoms which may be imagined to be contained in a regular sided box. The smallest such unit is called a unit cell [Fig. 6·4(a)]. If many such boxes are stacked together, all in the same orientation, like bricks [Fig. 6·4(b)], the atoms are automatically placed in their correct positions in the crystal. The example shown is diamond which may not look familiar as it has been drawn in a different orientation. This unit cell contains 12 atoms but in general the unit cell may contain more or less than this number and the cell may be other than cubic in shape. The formal definition of the unit cell is the smallest group of atoms which, by repeated translation in three dimensions, builds up the whole crystal.

Table 6·1   Ionic radii

Ion	Radius (Å)	Ion	Radius (Å)
Li^+	0·78	U^{4+}	1·05
Na^+	0·98	Cl^-	1·81
K^+	1·33	Br^-	1·96
Rb^+	1·49	I^-	2·20
Cu^+	0·96	O^{2-}	1·32
Be^{2+}	0·34		
Mg^{2+}	0·78	S^{2-}	1·74

6·3   Interatomic distances and ionic radii

It was mentioned in Chapter 5 that the closed electron shell which is created when an ionic or metallic bond is formed can be regarded roughly as a rigid sphere so that adjacent atoms in a solid pack together like solid balls. The radius of the equivalent rigid sphere is termed the ionic radius of the element, and some figures are given in Table 6·1. These have been deduced from the interatomic distances in ionic and metallic solids, measured using X-ray diffraction. It is also possible to calculate them from the diameter of the electronic wave-functions obtained by solving Schrödinger's equation. The agreement with measured values is reasonably good in view of the approximations which must be made in performing the complex calculations.

An obvious starting point for the study of crystal structures is therefore the consideration of the kinds of structures which can be built by packing identical spheres together as closely as possible. Later on structures composed of spheres of two different sizes will be discussed.

6·4   Close packed structures of identical spheres

Fig. 6·5(a) shows balls packed together on a plane surface. Each is surrounded by six neighbours which just touch it. If we place two such planes in contact each ball in the second layer rests on three from the lower layer. In Fig. 6·5(b) the centres of the spheres in the lower layer are marked A while those of the second layer are marked B. When we come to add a third layer, the centres of the spheres in it can be placed over the gaps in the second layer marked C or over those marked A. In the latter case they are directly above the spheres in the first layer. We show a side view of this arrangement in Fig. 6·5(c): the sequence of positioning the layers may be represented as A-B-A-B-A . . . etc. However, in the other case the third layer does not sit over the first so it is positioned differently from either layer A or B. Since the fourth layer must begin to repeat the sequence we obtain [Fig. 6·5(d)] the pattern A-B-C-A-B-C . . . etc.

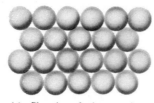

(a) Plan view of spheres resting on a plane surface

(b) A second layer of spheres (shown dashed) is placed on the first (plan view)

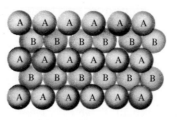

(c) Side elevation of several layers stacked A-B-A-B-A

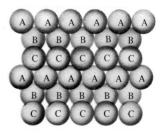

(d) Side elevation of several layers stacked A-B-C-A-B-C-A-

Fig. 6·5 The formation of the two close packed structures of spheres

These two closest packed arrangements are shown again in Fig. 6·6 where the relative atomic positions are seen more clearly and the unit cell is shown in each case. The two structures are known as *face-centred cubic* (f.c.c.) and *close-packed hexagonal* (c.p.h.) respectively. In the former structure the cube shown is not the smallest possible 'building brick' but it is usually referred to as the unit cell.

These close-packed structures should be preferred by solids with van der Waals or metallic bonding because the atoms or ions have closed shells and therefore may be treated as hard spheres. Many metals do indeed crystallize in the face-centred cubic structure and so do all the noble gases excepting only helium, which chooses the close-packed hexagonal structure. Among the metals the c.p.h. structure is exemplified by cobalt, zinc, and several others.

Since ionic solids do not normally consist of atoms of identical size they do not form these close-packed structures.* Similarly, covalent substances do not normally seek closest packing: the main consideration is that the direction of their bonds be maintained since these are rather rigid.

* See, however, the footnote on p. 83.

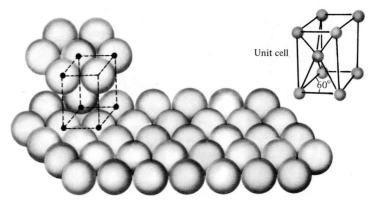

(a)   Close-packed hexagonal structure

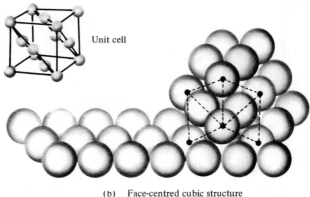

(b)   Face-centred cubic structure

Fig. 6·6 Isometric views of the c.p.h. and f.c.c. structures, showing the unit cells

These two are the only ways in which spheres can be closely packed together but there are some arrangements with looser packing which appear in nature. One of these is shown in Fig. 6·7, and is called body-centred cubic (b.c.c.). As can be seen by inspecting the atom in the centre of the cube, each atom has only eight nearest neighbours as compared to twelve in both of the close-packed structures. This is indicative of the loose packing, in spite of which many metals crystallize in the b.c.c. form. They include all the alkali metals and, most important, iron. The number of nearest neighbours, eight in this case, is called the *coordination number* of the structure.

The simplest cubic arrangement possible, also shown in Fig. 6·7, is called simple cubic. This is a very loosely packed structure and is not of much importance since it is exhibited

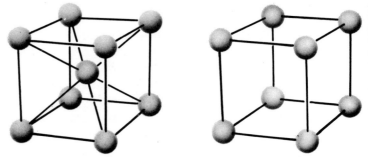

Body-centred cubic   Simple cubic

Fig. 6·7 The unit cells of the b.c.c. and s.c. structures

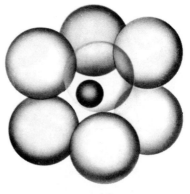

(a)

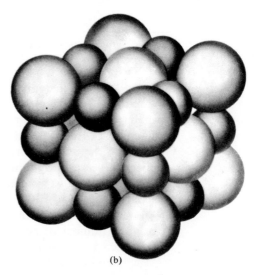

(b)

Fig. 6·8 (a) The CsCl unit cell, showing that if the central anion is too small it cannot touch the cations; (b) the rock salt unit cell

only by the metal polonium. Each atom has only six nearest neighbours.

6·5   Ionic crystals

It was mentioned earlier that the ratio of the ionic radii is important in determining the structure of an ionic solid. To understand this question it helps to study the problem in two dimensions by arranging coins on a table. With identical coins, it is possible to make each touch six neighbours. Now take a shilling (five new pence) and some florins (or ten new pence pieces) the ratio of whose radii is 0·83. It is possible to place five florins in contact with one shilling but one cannot make an extended lattice in this way. However, an arrangement with four nearest neighbours, making a square lattice, can be repeated indefinitely in every direction.

The three-dimensional case is illustrated in Fig. 6·8(a) where we show how a positive ion (a *cation*) placed in the body-centred position in a simple cubic structure of *anions* (negative ions) does not touch the anions if it is too small compared to them. Because of this, the forces of attraction and repulsion between the ions do not balance and the structure is not stable. It is easy to show (Problem 6·1) that the critical ratio of radii for which the b.c.c.-like structure is just stable is $r^+/r^- = 0·732$. If the ratio is smaller than this it is possible that the arrangement shown in Fig. 6·8(b) will be stable: with the same radius ratio as in Fig. 6·8(a) the cation, placed in the body-centred position in an f.c.c. arrangement of anions, just touches the six face-centred anions giving a stable arrangement. Figure 6·9(a) and Fig. 6·9(b) show the extended lattices corresponding to those in Fig. 6·8. We see that the lattice in Fig. 6·9(a) can be regarded as two interpenetrating simple cubic lattices, one for each kind of ion. This structure is called after the ionic compound caesium chloride (CsCl) which crystallizes in this way. Each ion is surrounded by eight nearest neighbours and eight is hence the coordination number of this structure.

We have met the structure in Fig. 6·9(b) before in NaCl and it takes the name of the natural crystalline form of this material, rock salt. It can be represented as two interpenetrating f.c.c. lattices, one of anions and the other of cations, shifted with respect to one another by half a cube edge.† Since each ion touches six neighbours the coordination number is six and the packing is yet looser than in the body-centred arrangement.

---

†Because of this, the structure is basically f.c.c., since it may be regarded as a single f.c.c. lattice in which each lattice point is associated with an Na–Cl atom pair. Note, however, that it is not close-packed in the sense of the previous section.

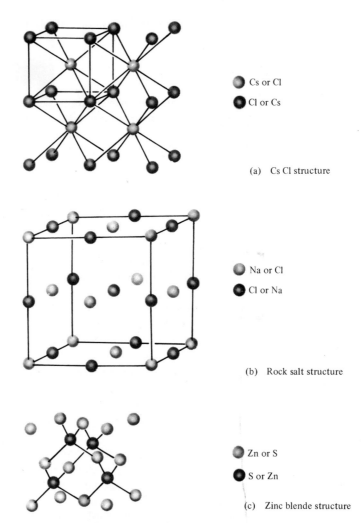

(a) Cs Cl structure

● Cs or Cl
● Cl or Cs

● Na or Cl
● Cl or Na

(b) Rock salt structure

● Zn or S
● S or Zn

(c) Zinc blende structure

Fig. 6·9 (a) Part of the extended lattice of CsCl, showing how it is built of unit cells; (b) the rock salt structure—a single unit cell is shown here; (c) the unit cell of the zinc blende structure

The coordination number is actually the same as that of the simple cubic lattice and indeed the two lattices are identical if one ignores the difference between the positive and negative ions.

It is clear that in the rock salt structure also there is a limit to how small the cation may be. This is easily found (Problem 6·2) to be 0·414 times the anion size.

For radius ratios below this we might expect the coordination number to reduce yet further. Since arrangements with five neighbours cannot fit together to give an extended space lattice, four is the next lowest coordination number. Once again we have met this already when we considered carbon—the diamond structure (Fig. 6·4). It should be emphasized that carbon itself has this structure because of the directional nature of its four covalent bonds and not because of its radius ratio (its radius ratio is unity).

The ionic relative of the diamond structure is shown in Fig. 6·9(c). Like rock salt and caesium chloride it can be represented as two interpenetrating lattices, both f.c.c. in this case. It is named after zincblende (ZnS) which itself is not wholly ionic but which is the commonest compound which crystallizes in this arrangement.

As an example of a predominantly ionic solid with the zincblende structure we may take beryllium oxide (BeO) although the bonds here are probably also partly covalent. However, this structure is favoured for the ionic bond since the radii of the two ions are 0·31 Å and 1·70 Å, that is, the radius ratio is 0·222.

It should be noted, however, that the radius ratio acts only as a rough guide to the structure of an ionic compound. There are many crystals which do not follow the rule, like lithium iodide (LiI) with a radius ratio of 0·28 which takes the rock salt structure in spite of the calculated radius ratio limit of 0·414. At the other end of the scale, potassium fluoride, radius ratio 0·98, also crystallizes in the rock salt structure. This is probably because the difference between the stabilities of the caesium chloride and sodium chloride structures is rather small and minor effects such as van der Waals forces can just tip the balance.

To summarize, we present the rules which govern the crystal structures of ionic solids, and finally in Table 6·2 we show how well (or how badly) they are followed.

(i) The non-directional ionic bond favours closest possible packing consistent with geometrical considerations.

Table 6·2   Ionic structures

Compound	Radius ratio	Predicted structure	Observed structure	Coordination number
NaCl	0·52	NaCl	NaCl	6
AgF	0·93	CsCl	NaCl	6
KBr	0·68	NaCl	NaCl	6
RbCl	0·82	CsCl	NaCl	6
CsCl	0·93	CsCl	CsCl	8
CsBr	0·87	CsCl	CsCl	8
RbCl	0·82	CsCl	CsCl†	8
BeO‡	0·22	ZnS	ZnS‡	4
BeS‡	0·17	ZnS	ZnS‡	4
CuF‡	0·71	ZnS	ZnS‡	4
LiI‡	0·28	ZnS	NaCl	6

†Structure at high pressure.
‡Partially covalent bonding.

(ii) If anion and cation cannot touch one another in a given arrangement, it is generally not stable.

(iii) The radius ratio of the ions roughly determines the most likely crystal structure.

6·6 Covalent crystals

It may be thought that the previous section deals with solids of no practical interest. Unfortunately, because of the mixed ionic-covalent bonding in so many solids, the underlying principles of ionic bonding are only clearly exemplified in the alkali halides. Among the covalent solids, however, we have a number of practical importance, mostly in the electronics industry. Carbon, silicon, and germanium all crystallize in the diamond structure as we have already seen. The main requirement in these covalent structures is that the direction of the bonds be appropriate to those of the wave-functions of the bonding electrons. The compounds between Group III atoms and Group V atoms (e.g., GaAs and InSb) form the related zincblende structure. This retains the fourfold coordination of diamond with the nearest neighbours at the corners of an imaginary tetrahedron—an arrangement called tetrahedral bonding. The preference for this structure has nothing to do with atomic radii but is determined by the directionality of the covalent bonds.

Several compounds between Group II and Group VI elements also choose this structure, e.g., ZnS (an obvious case) and BeS, while others like CdS, CdSe are normally found in a modified form of the zincblende structure called wurtzite. Naturally there is more than an element of ionic bonding present in these compounds. In other sulphides, notably PbS and MgS, the ionic nature of the bond predominates and the rock salt structure results. This is also true of many oxides of Group II elements such as MgO, CaO, although BeO as we saw takes the zincblende structure on account of the smallness of the Be ion. Among the covalent oxides is ZnO which has the wurtzite structure and CuO, PdO, and PtO which all form yet another kind of structure with fourfold coordination.

Because oxygen has a lower atomic number than sulphur it is more electronegative and hence tends to form ionic bonds more readily. Thus the covalent oxides mentioned above are in the minority.

6·7 Crystals with mixed bonding

We also find some covalently bonded solids with a marked metallic character. Although not really in this category, graphite (an allotrope of carbon) conducts electricity and is a soft material quite unlike diamond. The structure (Fig. 6·10)

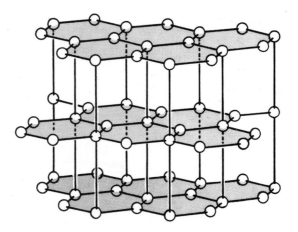

Fig. 6·10 The structure of graphite

explains this: it consists of sheets of covalently bonded carbon atoms with threefold coordination, loosely bonded by van der Waals forces to neighbouring sheets. The large distance between sheets is a reflection of the weakness of the van der Waals bond and this allows the sheets to slide over one another rather readily. This gives graphite its lubricating property, which is of such great importance since it is one of the few lubricants capable of withstanding high temperatures without vaporizing. In addition there is one free electron per atom since only three covalent bonds are formed. These free electrons cannot easily cross from sheet to sheet because of the large distance involved, so the conducting property is directional, and the resistance is low only along the sheets of atoms.

Although graphite is not strictly metallic, antimony and bismuth certainly are, in spite of their position in Group VB of the Periodic Table. Their structure is related to that of graphite, although the sheets of atoms are not flat, but the important difference is that the distance between sheets is much smaller. This is because the inter-sheet bonding is predominantly metallic and therefore stronger. As a result the conductivity is much higher and less directional than in graphite. Note that the coordination number is much lower than is expected of a metal because of the covalent bonds. However, in liquid form this influence of the covalent bonds is lost and the atoms pack more closely together, so that Sb and Bi both contract on melting. It is this property which makes them important in typecasting where the expansion as they solidify enables them to fill all the interstices in a mould.

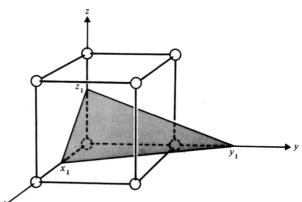

Fig. 6·11 Construction to determine the Miller indices of a plane (shown shaded) in a cubic lattice

6·8 Atomic planes and directions

In later chapters we shall often wish to refer to sets of planes, as for instance the cube faces in the f.c.c. structure, shown again in Fig. 6·11, or the sheets of atoms in the graphite structure. Referring to Fig. 6·11, we set up x, y, and z axes first, along the cube edges. Suppose that we wish to describe a plane, such as that shaded in Fig. 6·11, which intersects the x, y, and z axes at distances x, y, and z from the origin. First take the reciprocals of these distances, $1/x$, $1/y$, and $1/z$. In general these will be fractions, so we clear fractions by multiplying them by a suitable factor, s, so that the resulting integers s/x, s/y, and s/z are the smallest integers which can be so formed. These are then termed the *Miller indices of the plane*.

For instance, suppose the intercepts were $0·5a$, $0·25a$, and $1·5a$ (a being the length of the cube edge as shown) then the reciprocals are $2/a$, $4/a$, and $2/3a$ respectively. Multiplying these by $3a/2$ gives the lowest set of integers, 3, 6, and 1 which are in the same ratio as the reciprocals. The plane is then called the (361) plane, these numbers being the Miller indices.

If any intercept is negative, the corresponding Miller index is also negative and this is denoted by placing a bar above it. So the plane with intercepts $-0·5a$, $0·25a$, and $1·5a$ is denoted $(\bar{3}61)$.

If a plane is parallel to an axis it is taken to intercept that axis at infinity. Thus the reciprocal is $1/\infty = 0$ and the appropriate Miller index is zero.

Example. The cube face parallel to both the y and z axes has intercepts 1, ∞, ∞, [Fig. 6·12(a)] and hence is denoted (100). The plane having equal intercepts on all three axes is denoted by (111) [Fig. 6·12(b)] and that containing two face diagonals [Fig. 6·12(c)] is denoted (110). Note that there are many planes

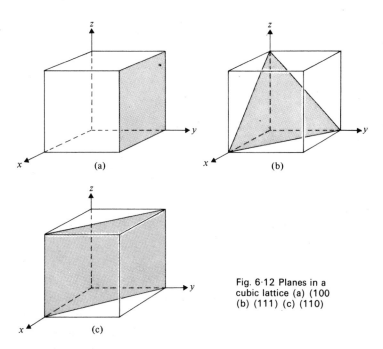

Fig. 6·12 Planes in a cubic lattice (a) (100 (b) (111) (c) (110)

of the same character with different indices, for example, (110), (101), (011), ($\bar{1}$10), ($\bar{1}$0$\bar{1}$), etc.

Since one often needs to refer to such a family of planes a notation has been adopted as follows: {100} denotes a family of planes of the type (100) and would include all those listed in the previous paragraph. Similarly, there are families of {110} planes and {111} planes. These three are the types we shall meet most frequently in this book.

6·9   Crystallographic directions

The direction of a line of atoms may be described using a similar notation. Here we take a parallel line which passes through the origin (Fig. 6·13) and note the length of the projections of this line on the x, y, and z axes; say they are x_0, y_0, and z_0. Then we reduce these to the smallest integers which bear the same ratio to one another. So if the intercepts were $0.75a$, $0.25a$, and a, the indices of the direction are 3, 1, and 4 since the relation $3:1:4$ is the same as $0.75:0.25:1$. This direction is denoted the [314] direction—note the square brackets to distinguish it from the (314) plane.

Just like the principal planes of importance, the directions with which we shall be primarily concerned are [110], [100], and [111]. These are, respectively, a cube face diagonal, a cube edge, and a body diagonal. Note that the fact that a plane and its normal have the same indices is a peculiarity of the cubic lattice and does not apply to other lattice types. Families of directions

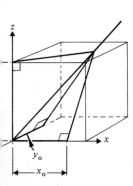

Fig. 6·13 Construction to determine the Miller indices of a direction in cubic lattice

are labelled by special brackets as are families of planes. Thus $\langle 100 \rangle$ denotes a family of directions which includes [100], [010], [001], [$\bar{1}$00], [0$\bar{1}$0], and [00$\bar{1}$].

6·10 Theory of crystal structures

In this chapter the atomic arrangements, or lattices, have been discussed only in a qualitative way. An alternative approach is to study them in a formal, abstract way just as if they are geometrical structures. If this is done one finds that only a limited number of arrangements is possible. We have already met one of the reasons for this when we considered the possibility of fivefold coordination. It is just not possible to build a regular, extended lattice with such an arrangement. Similarly, many other polyhedra (that is, arrangements with different coordination numbers) cannot be stacked to form space lattices. There is another example of the packing of coins which demonstrates this: the reader may care to try building a two-dimensional lattice with the seven-sided 50 new pence piece. (He is advised not to spend too long trying.) In this kind of analogy the seven-sided coin represents the unit cell of the lattice. As a result of such considerations Bravais discovered that there are only 14 different kinds of unit cell which can form an extended lattice. They are known as Bravais lattices, after their discoverer, and we show them in Fig. 6·14. Each lattice point (represented by a circle in the figure) may accommodate more than a single atom and in many crystals it does so. In consequence there may be many more than just fourteen kinds of crystal structure. These variations may be classified according to the degree of symmetry exhibited by the group of atoms situated at each lattice point. Since there is a limited number of kinds of symmetry the total number of crystal classes is restricted to 230.

It is important to note that each of these 230 classes may be realized in many different ways using different atoms or molecules so that there is virtually no restriction on the range of different materials which may be made.

6·11 Measurements on crystals

Earlier in the chapter it was mentioned that the distances between atomic planes can be measured by using the diffraction of X-rays. The basic requirements are (i) a narrow beam of X-rays of known wavelength, (ii) a means of recording the direction of the diffracted beam (usually a photographic film or plate), and (iii) a turntable on which the crystal is mounted so that it may be suitably oriented with respect to the beam. These elements are shown in Fig. 6·15 but the details are beyond

Fig. 6·14 The fourteen Bravais lattices

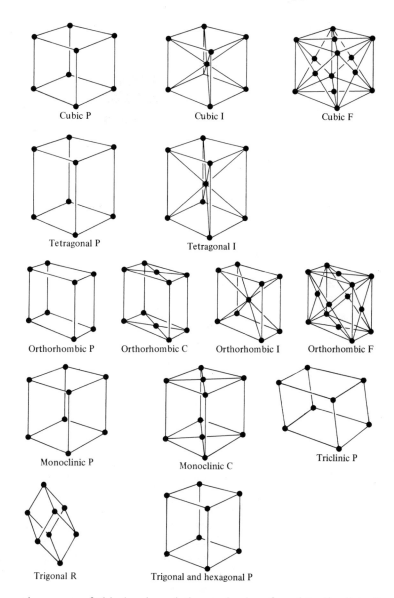

Cubic P   Cubic I   Cubic F

Tetragonal P   Tetragonal I

Orthorhombic P   Orthorhombic C   Orthorhombic I   Orthorhombic F

Monoclinic P   Monoclinic C   Triclinic P

Trigonal R   Trigonal and hexagonal P

the scope of this book and the reader is referred to the list of further reading at the end of the book. In this way the dimensions of the unit cell of every common material have been measured. The dimensions quoted in reference books are usually the lengths of the edges of the unit cell, and these are called the *lattice constants*. Naturally, in a cubic material, only one such constant is required to specify the size of the cell.

*6·12 Bragg's law

We conclude with a derivation of the law which relates the wavelength of an X-ray to the spacing of the atomic planes—

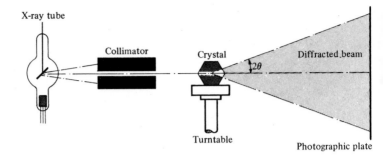

Fig. 6·15 Apparatus for the determination of crystal structure by X-ray diffraction

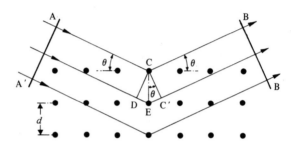

Fig. 6·16 The proof of Bragg's law

Bragg's law. We begin by assuming that each plane of atoms partially reflects the wave as might a half-silvered mirror. Thus the diffracted wave shown in Fig. 6·16 is assumed to make the same angle, θ, with the atomic planes as does the incident wave.

Now the criterion for the existence of the diffracted wave is that the reflected rays should all be in phase across a wavefront such as BB′. For this to be so, the path lengths between AA′ and BB′ for the two rays shown must differ by exactly an integral number n of wavelengths λ.

Thus δ, the path difference, is given by $\delta = n\lambda$, $(n = 1, 2, 3 \ldots)$. Now since lines CC′ and CD are also wavefronts we have

$$\delta = 2EC'$$
$$= 2CE \sin \theta$$
$$= 2d \sin \theta$$

Thus

$$2d \sin \theta = n\lambda \ (n = 1, 2, 3 \ldots)$$

This is just Bragg's law.

The assumption that each atomic plane reflects like a mirror may be justified by a similar argument to that above assuming that each atom in the plane is the source of a secondary

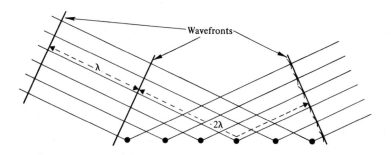

Fig. 6·17 Reflection of X-rays from a plane of atoms

Huyghen's wavelet (Fig. 6·17). The scattered waves all add up in phase as shown to form the reflected ray.

Problems

6·1 In the CsCl structure shown in Fig. 6·10(a) let the ionic radii be r_1 and r_2 ($r_2 > r_1$). Assuming that the anions just touch the cations, calculate the length of the body diagonal, and hence derive the value of the radius ratio at which the structure just becomes unstable.

Using the ionic radii listed in Table 6·1, find the size of the largest impurity ion which can be accommodated interstitially (i.e., between the host ions) in the CsCl lattice with its centre at the point $(\frac{1}{2}, \frac{1}{2}, 0)$ in the unit cell.

6·2 If the radii of the ions in the NaCl structure are r_1 and r_2 ($r_2 > r_1$), calculate the length of the face diagonal in the unit cell and hence show that the structure is stable only when $r_1 > 0·414 \, r_2$.

6·3 What crystal structure do you expect to find in the following solids: BaO, RbBr, SiC, GaAs, Cu, NH_3, BN, SnTe, Ni? (Use the ionic radii in Table 6·1; the radius of Ba^{2+} is 1·35 Å.)

6·4 Calculate the angles for first-order diffraction (i.e., $n = 1$) from the (100) and (110) planes of a simple cubic lattice of side 3 Å when the wavelength is 1·0 Å. If the lattice were b.c.c., would you expect to find diffracted beams at the same angles?

6·5 How many atoms are there in the unit cell of (a) the b.c.c. lattice (b) the f.c.c. lattice.

6·6 Determine the Miller indices of a plane containing three atoms which are nearest neighbours in the diamond lattice. [Use Fig. 6·4 (a).]

6·7 Calculate the density of graphite if the interplanar distance is 3·4 Å and the interatomic distance within the plane is 1·4 Å. (Note: density = mass of unit cell ÷ volume of unit cell.)

6·8 The density of solid copper is $8·9 \times 10^3$ kg/m³. Calculate the number of atoms per cubic metre.

In an X-ray diffraction experiment the unit cell of copper is found to be face-centred cubic, and the lattice constant (the length of an edge of the unit cell cube) is 3·61 Å. Deduce another figure for the number of atoms per cubic metre and compare the two results. What factors might give rise to a discrepancy between them?

6·9 Place the materials listed in Problem 6·3 in the expected order of increasing melting points. Compare with the melting points given in a standard reference book.

6·10 The lattice constants of the metals Na, K, Cu, and Ag are respectively 4·24 Å, 4·62 Å, 3·61 Å, and 4·08 Å. Both Na and K have b.c.c. structures, while Cu and Ag are f.c.c. Calculate in each case the distance between the centres of neighbouring atoms, and hence deduce the atomic radius of each metal.

6·11 What are the Miller indices of the close-packed planes of atoms in the f.c.c. lattice (see Fig. 6·6)?

7 Interatomic forces and crystal defects

7·1 Introduction

In the earlier chapters of this book great stress was laid on the quantum-mechanical aspects of the electrons in an atom. In particular the idea of the particulate electron confined to a specific orbit was shown to be inadequate and the electron distribution around the atom was best thought of in terms of charge clouds. Nevertheless, when considering crystal structures, especially of metals, we have seen that the arrangements of the atoms are consistent with picturing them as being 'hard' spheres, closely packed together. This is justified quantum-mechanically by (a) the spherical symmetry of closed electron shells and (b) the repulsive force between overlapping electron clouds, which rises very rapidly as the overlap increases. In the following chapters on mechanical and thermal properties we will make the assumption that no serious error is involved in treating the atoms as hard spheres.

7·2 Interatomic forces and elastic behaviour

We have seen in Chapter 5 that various mechanisms of bonding exist for holding atoms together in the solid state. All these result in a net attraction between the atoms and at a certain distance apart this attraction is balanced by a repulsion due to overlap of the electronic wave functions or charge clouds.

We can express this behaviour for a pair of atoms by assigning to them a potential energy, V, which is a function of distance, r, between them. Thus we write

$$V = \frac{-A}{r^n} + \frac{B}{r^m} \tag{7·1}$$

where, following the usual convention for potential energy, the term arising from an *attractive* force is given a negative sign and that for a *repulsive* force a positive sign. In this expression A and B are constants and n and m are to be determined; in the case of electrostatic attraction between unlike charges, for example, n would be unity. Again, by definition, the net force between the atoms will be given by

$$F = \frac{-\partial V}{\partial r} = \frac{-nA}{r^{n+1}} + \frac{mB}{r^{m+1}} \tag{7·2}$$

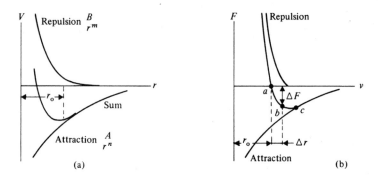

Fig. 7·1 The Condon–Morse curves; (a) potential energy V and (b) force F as a function of interatomic spacing r

and, if the atoms are in equilibrium, this force will be zero at some critical distance r_0, that is

$$0 = \frac{-nA}{r_0^{n+1}} + \frac{mB}{r_0^{m+1}}$$

These equations are illustrated graphically for arbitrary values of n and m in Fig. 7·1: these are known as the Condon–Morse curves after their originators. While Fig. 7·1 is derived for an isolated pair of atoms, similar behaviour will obviously occur between each pair of atoms in a crystal lattice. The actual values of n and m will depend on the nature of the crystal bonding forces and on the crystal itself. They have, for example, been calculated for an ionic crystal such as rock salt, in which n is one, as would be expected for electrostatic attraction, and m is in the region of ten.

When an external force is applied to the crystal such as to pull the atoms further apart they will be displaced from their equilibrium positions until the applied force is balanced out by the increase, ΔF, in attractive force between the atoms corresponding to moving from point a to b in the curve of Fig. 7·1(b). Suppose that the applied force elongates the crystal from length l to $l+\Delta l$; then the fractional change in length produced is $\Delta l/l$. The displacement of each atom from its equilibrium position will be in just the same proportions: thus

$$\frac{\Delta r}{r_0} = \frac{\Delta l}{l}$$

If the crystal behaves elastically the atoms will return to their original positions when the external force is removed.

The ratio $\Delta F/\Delta r$ will, for small values of Δr, be given by the slope of the tangent to the curve at the equilibrium position, that is by $(\partial F/\partial r)_{r=r_0}$. In practice elastic behaviour in crystals rarely produces fractional elongations in excess of $\frac{1}{2}\%$ so that Δr will always be small and the slope $\partial F/\partial r$ of the force curve can be regarded as a constant over this small range of r. This means that over the elastic range the displacement, Δr, of the atoms is proportional to the external applied force in accordance with Hooke's law of elasticity.

Precisely the same arguments apply if a compressive external force is applied; the equilibrium distance between the atoms is reduced until the increased repulsive force between the atoms balances the applied force.

If we consider a very large external tensile force applied so that the equilibrium distance increases to the point c in Fig. 7.1(b), the material will completely fracture since this point corresponds to the cohesive force of the material. From the curve it will be seen that the fractional charge $\Delta r/r_0$ is still quite small at the point c. Thus the material will have fractured after only quite a small elastic elongation. This behaviour is, in fact, observed in what are termed brittle materials such as a pure silicon or germanium crystal for example. Most metals and many other materials, however, exhibit considerable elongations—up to 60% in the case of copper—before fracture occurs. Furthermore, these materials do not return to their original shape after the application of large forces. Clearly the simple picture based on interatomic binding forces is not adequate to explain the mechanical properties of materials. In fact mechanical properties are largely determined by the defects which exist in the regular crystalline structure of all materials and we must, therefore, consider the nature and influence of these. First, however, the types of mechanical stress and their effects will be reviewed.

7·3   Elastic and plastic behaviour

The simplest mechanical test is a tensile test in which a load is applied to the specimen to elongate it at a constant rate. The loads necessary to produce given elongations are recorded as the test proceeds. The load is measured as a *stress*, σ, which is defined as the load per unit cross-sectional area on the sample. Thus, if the load is P kgf and the cross-sectional area is $A\,\mathrm{m}^2$ then the stress is given by $\sigma = P/A\,\mathrm{kgf/m}^2$. The elongation is measured in terms of *strain*, ε, which is defined as the fractional change in length of the specimen. Thus if the sample is elongated from l to $l+\Delta l$ metres, the strain is given by $\varepsilon = \Delta l/l$ and is a dimensionless number.

Initially, during the test, the material elongates elastically and obeys Hooke's law which states that

$$\sigma = E\varepsilon \qquad\qquad\qquad (7\cdot3)$$

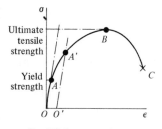

Fig. 7·2 Stress–strain diagram for copper

where E is a constant for the material and has the same dimensions as stress; it is called *Young's modulus*.

As the specimen elongates in the elastic region its cross-sectional area decreases. If we assume our test specimen has a square cross section, each side being d metres, then the transverse strain will be $-\Delta d/d = \varepsilon_t$ and can be measured as the test proceeds. This is related to the tensile strain ε by *Poisson's ratio*, ν, which is also a property of the material. It is defined by

$$\nu = \frac{\varepsilon_t}{\varepsilon} \tag{7·4}$$

and is dimensionless.

The results of a tensile test are presented in the form of a stress–strain diagram and a typical one for copper is shown in Fig. 7.2. The section OA corresponds to the elastic region, which means that if the load is removed at any point the line OA is retraced. Beyond the *elastic limit* at the point A *plastic* deformation occurs and the specimen undergoes a permanent change in shape. If the specimen is unloaded at the point A' the line A' O' is followed; on reapplication of the load the specimen retraces the line O' A' and, in general, exhibits elastic behaviour over the whole of the region O' A'. Thus plastic deformation appears to increase the range of elastic behaviour in the deformed specimen. This is called *work hardening* or simply *cold work*. At the point B the ultimate tensile strength is reached and the specimen continues to elongate, even with a reduction in the applied stress, until fracture occurs at the point C.

While the tensile test is the most straightforward to visualize and understand, there are many other ways in which force may be applied to a solid to produce deformation. These may always be reduced to a combination of three types of force: (a) uniaxial tensile or compressive, (b) shear, and (c) bulk compressive or tensile.

Shear can be understood by reference to Fig. 7·3. We suppose that a cube of side l is cemented firmly to a base and that a force is applied along the upper surface, parallel to it, producing the deformation shown in the diagram. The shear stress, τ, is the force applied divided by the area of the plane in which the force is acting. The shear strain, γ, defined as $\tan \theta$ in the above diagram, is given by

$$\gamma = \delta/l$$

The shear stress–shear strain curve exhibits the same features as the tensile one, that is, an elastic region and a plastic region followed by ultimate fracture. Over the elastic region we define

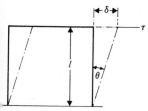

Fig. 7·3 Illustrating shear stress and strain

the shear modulus, G, by

$$\tau = G\gamma \tag{7·6}$$

in accordance with Hooke's law.

If a hydrostatic pressure, P, is applied to a specimen the strain produced is due to a change in volume. Again elastic and plastic behaviour may be exhibited and over the elastic region we define a *bulk modulus*, K, by

$$P = K.(\Delta V/V) \tag{7·7}$$

where $\Delta V/V$ is the volume strain produced.

Under normal working conditions a mechanical component is designed to behave elastically under all the forces to which it is subjected in use. In initial manufacture, however, the material undergoes a permanent change of shape and this will generally involve plastic flow. Thus the study of both elastic and plastic behaviour of materials is of importance.

7·4 Elastic energy

When the atomic bonds of a material are strained by the application of an external force elastic energy is stored in the body and can be fully recovered as mechanical work done by the body as the applied force is reduced to zero. This strain energy can be calculated by considering a cube with sides of length, l_0, strained in uniaxial tension by a force which rises from zero to some value, F. The loaded faces of the cube move apart a distance δl and the force does work given by force $\times$ distance. But for a solid obeying Hooke's law the extension, δl, is proportional to the applied force, F, that is, when the force is zero the extension is zero. Thus the work done in reaching the extension δl is just $\frac{1}{2}F.\delta l$.

Now, by definition, the tensile stress $\sigma = F/l_0{}^2$ and the strain $= \delta l/l_0$. Thus the work done per unit volume is given by $\frac{1}{2}F\delta l/l_0{}^3$ since the volume of the cube is $l_0{}^3$, and this is equal to $\frac{1}{2}(\sigma l_0{}^2 . \delta l/l_0{}^3)$ which equals $\frac{1}{2}\sigma\varepsilon$.

Thus

$$\text{strain energy density} = \tfrac{1}{2}\sigma\varepsilon = \frac{\sigma^2}{2E} = \tfrac{1}{2}\varepsilon^2 E \tag{7·8}$$

by use of Eqn (7·3).

Similarly for shear stress, using Eqn (7·5),

$$\text{strain energy density} = \tfrac{1}{2}\tau\gamma = \frac{\tau^2}{2G} = \tfrac{1}{2}\gamma^2 G \tag{7·9}$$

For hydrostatic deformation

$$\text{strain energy density} = \tfrac{1}{2}P\Theta = \frac{P^2}{2K} = \Theta^2 K \qquad (7\cdot10)$$

where $\Theta = \Delta V/V$ in Eqn (7·7).

The density of strain energy stored when a body is on the point of failure is called the *modulus of elastic resilience*. This is about $2\cdot5\times10^6$ J/m³ for strong steel which works out at about 2×10^{-4} eV per atomic bond. Compared with bond energies which are of the order of two to four electron volts it is seen that the strain energy stored is extremely small on an atomic scale.

7·5 Slip planes

Earlier in this chapter the atomic basis of the strength of materials was discussed and it was stated that very large forces are required literally to pull the atoms apart to fracture the material. In a real three-dimensional crystal it is often not necessary to exceed the bonding forces in order to produce a permanent change of shape, that is, plastic flow.

Consider the diagram of Fig. 7·4(a) in which a speciment is shown loaded as in a tensile test. If we cut the specimen along the line xx′ and applied the same tensile force, P, it may be resolved into components P_S and P_T, as shown in Fig. 7·4(b). Now P_S is clearly a shear force along the plane xx′ while P_T is a tensile component perpendicular to the plane. Thus even a simple tensile force will produce shear stresses in the sample. Now let us show the sample as a set of close-packed spherical atoms as in Fig. 7·4(c). In the direction xx′ the atoms are in close-packed planes and it will be a relatively simple matter for one of these planes to slide over the other, under the influence of the shear stress. This is called *slip*, and the plane xx′ is a *slip plane*. As one plane of atoms slip over another, the atoms in the adjacent planes obviously must move apart, but then they drop back to new close-packed positions, as in Fig. 7·4(d). The

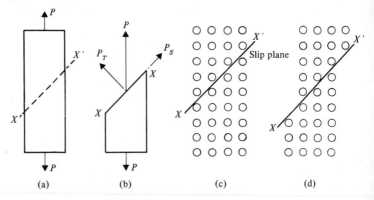

(a)   (b)   (c)   (d)

Fig. 7·4 Resolution of tensile stress and strain (see text)

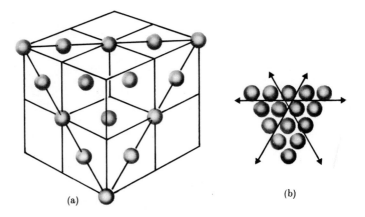

Fig. 7·5 (a) the close packed (111) plane and (b) slip directions in a (111) plane and in a face-centred cubic crystal

energy or force to bring this about will obviously be much less than that involved in actually breaking the bonds between atoms. In general, where the crystalline form of the material is such as to allow a number of slip planes the crystal will be much weaker in shear than in tension and will, therefore, yield by slip giving plastic flow. Direct evidence for this behaviour is obtained by testing a single crystal in tension; the steps on the surface resulting from slip are clearly seen.

Slip generally occurs on planes of closest packing, and this gives a rough basis for estimating the ductility to be expected of various crystal structures.

a. F.C.C.

In the face-centred cubic crystal the close-packed planes are {111} illustrated in Fig. 7·5(a), and there are four such planes in the structure. In each plane there are preferred slip *directions* which correspond to the grooves between the lines of atoms. There are three preferred directions of the type [110], as shown in Fig. 7·5(b) in each of the four planes so that there are, altogether, 12 possible slip systems in the f.c.c. structure. A photograph of slip bands in copper, which is a f.c.c. metal, can be taken if they are delineated by etching the surface of the specimen with a chemical which preferentially dissolves the disordered regions between the slipped sections.

b. B.C.C.

There are actually no close-packed planes in the body-centred cubic structure. The most closely packed planes are {110} as illustrated in Fig. 7·6, of which there are six, and the preferred directions in this plane are of the type ⟨111⟩ of which there are

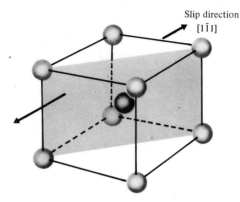

Fig. 7·6 The (110) plane in a body-centred cubic crystal

two in the plane, giving 12 slip systems. However, there are two other planes {211} and {321} which each contain one ⟨$\bar{1}$11⟩ direction in which slip is possible and, altogether, as many as 48 slip systems can be found in this structure.

c. C.P.H.

The close-packed hexagonal structure has close-packed planes parallel to the basal plane. In each of these there are three possible slip directions so that these are three principal slip systems. The side of the hexagon also contains one of the slip directions and three of the six sides correspond to the three different slip directions giving a further three subsidiary slip systems. There is, in fact, a total of twelve possible systems.

In general, ductility in metals decreases in the order:

f.c.c. (Ag, Au, Cu, Al) → b.c.c. (Fe, Cr, Mn, Mo) → c.p.h. (Be, Zr, Zn)

From knowledge of the bonding energies between atoms it is possible to estimate the order of shear stresses necessary to cause plastic flow by the above mechanisms. Experimental measurements give values which may be over one thousand times smaller than those predicted. The resolution of this problem was found in the dislocation theory.

7·5 Dislocations

An extra half-plane of atoms inserted between the parallel planes of atoms in a normal crystal can easily distort the crystal lattice so that it is accommodated, as shown in Fig. 7·7. This is called an *edge dislocation* and is able to move through the crystal, causing slip, under a shear force much less than that necessary to cause slip of a plane as a whole. One can see this in a general way by referring to Fig. 7·8. Initially the atoms in the

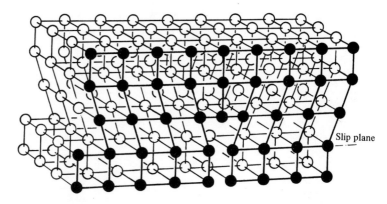

Fig. 7·7 An edge dislocation

planes adjacent to the edge dislocation are in a state of strain; in particular the pairs 1–2 and 3–4 are at a distance apart greater than their normal equilibrium distance and so are storing a certain amount of strain energy. When the shear stress, τ, is applied 1 moves to 1', decreasing the strain energy between 1 and 2. The edge dislocation atom 5 will move to 5', approaching more nearly the normal interatom equilibrium distance from 4, while 3, moving to 3', is being carried further away from 4 and increasing the strain energy. This increase is offset by atom 3 becoming the edge dislocation atom, while 1–2 and 5–4 become pairs in adjacent planes. The dislocation has thus moved, under the influence of the shear stress, a distance of one lattice constant to the right. If it continues to move the final result will be slip of the whole plane containing the edge dislocation by this amount, as illustrated in Fig. 7·9. It should be noted that the plane of the edge dislocation and its direction of movement are perpendicular to the slip plane.

A second type, known as the *screw dislocation*, is illustrated in Fig. 7·10. In this the atoms are displaced in two separate planes perpendicular to each other. Once again, under the

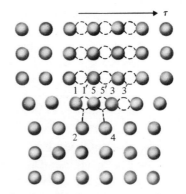

Fig. 7·8 Movement of atoms under shear stress around an edge dislocation

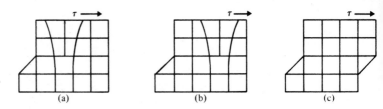

Fig. 7·9 Progression of an edge dislocation through a crystal to produce slip

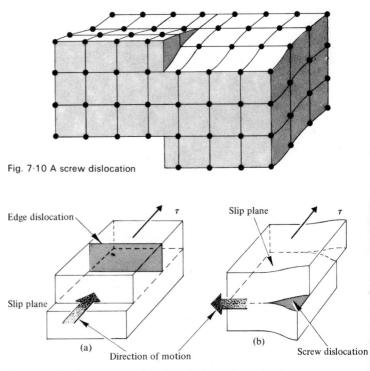

Fig. 7·10 A screw dislocation

Fig. 7·11 Directions of slip and motion of (a) an edge and (b) a screw dislocation

influence of shear stress the dislocation moves in the slip plane but in this case the direction of motion of the dislocation is perpendicular to the direction of slip. The two cases are compared in Fig. 7·11.

In general, dislocations are more complicated than the simple ones described above but they can always be resolved into a combination of edge and screw components.

7·7 Burgers' vector

The magnitude and direction of the slip associated with a dislocation is given by the Burgers' vector, **b**. If we take a circuit around a dislocation as shown in Fig. 7·12 and require that

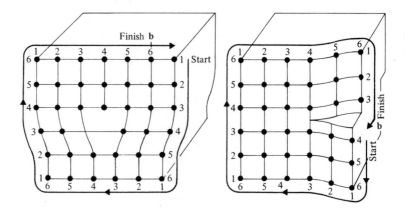

Fig. 7·12 The Burgers' circuit defining the Burgers' vector for (a) an edge
and (b) a screw dislocation

each parallel pair of 'legs' of the circuit have on them the same
number of atoms then the path will fail to close by the Burgers'
vector. Reference to the diagram shows that the direction of the
vector is just the direction of slip produced as the dislocation
moves. This is taken as positive when directed towards the
unslipped portion of crystal, that is, taking the Burgers' circuit
clockwise from the unslipped to slipped portions gives the
positive direction of the Burgers' vector. Since the lattice for a
perfect dislocation must be in registry across both the slipped
and unslipped regions, the Burgers' vector will always be an
integral of the atomic spacing for such a dislocation.

The force on a dislocation, which causes it to move when
stress is applied, is simply related to the Burgers' vector.
Referring to Fig. 7·13, in which the symbol $\perp$ represents an
edge dislocation, we assume a crystal of unit thickness perpendicular to the direction of slip to have the shear stress, τ,
applied as shown. The edge dislocation moves under the influence of the force, F per unit length of dislocation, which is
parallel to the stress τ and the work done in moving unit length
of the dislocation through the crystal will be given by
$W = FL$. Now the force due to the stress will be $\tau \times \text{area} =
\tau \times L \times 1$ and this will do work in producing a slip of b in a
direction parallel to F given by τLb. This must equal the work
done in moving the dislocation so that we have, for the force
per unit length of dislocation

$$F = \tau b$$

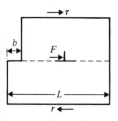

Fig. 7·13 Illustrating the
force F on an edge
dislocation for a stress,
τ, and Burgers' vector b

7·8   Climb

The movement of a dislocation along a slip plane is usually
referred to as *glide* and the gliding may be stopped by various

types of obstacle. In particular a grain boundary, where the atoms are not in correct registry, will effectively prevent the dislocation from moving further. The effects of this and other types of obstacle will be discussed in more detail in Chapter 8. An edge dislocation may succeed in moving in its *own* plane, that is, perpendicular to the slip direction, if atoms or vacancies (sites of missing atoms) are able to diffuse through the crystal. Clearly, if a line of extra atoms can move into the plane of the edge dislocation and add themselves on to the edge of the extra half-plane then the dislocation will have moved down by one lattice constant into the slip plane below. Similarly, if a line of vacancies can move on to the end of the half-plane the edge dislocation moves up by one lattice constant into the slip plane above. This is called *dislocation climb* and may operate to overcome an obstacle in the slip plane so that the dislocation climbs into another plane and then continues to glide. This will be assisted by elevated temperatures which speed up the rate of diffusion of interstitial atoms and of vacancies.

7·9 Jogs

The dislocation climb described above required a complete line of atoms to be added to, or taken away from, the edge dislocation over the whole of its length. This may not happen and the extra half-plane may be extended or diminished over only a portion of its length. In this case the dislocation will be operating on two slip planes simultaneously and the point where it moves from one slip plane to the next is called a *dislocation jog*. An edge dislocation may obviously have a number of jogs along its length.

Whilst it is more difficult to visualize, a screw dislocation may also contain jogs. The difference in this case is that for the screw the slip direction is that of the dislocation line. The jog can be thought of as a small piece of edge dislocation inserted into the screw dislocation. The direction for easy movement of this edge component is perpendicular to the direction of motion of the screw dislocation itself so that it tries to move itself across the screw dislocation. A jogged screw dislocation can, nevertheless, continue to move in the slip plane carrying the jog along with it. The edge dislocation cannot move in the same direction as the screw dislocation without atoms diffusing to it to extend the half-plane with the result that as it moves it leaves a trail of vacancies behind it.

The presence of jogs makes dislocations less mobile and a greater amount of work is necessary to move a jogged dislocation than a straight one.

Fig. 7·14 Illustrating the generation of a stacking fault (plan view)

7·10 Partial dislocations

It has been pointed out that when the crystal lattice is in registry across the slipped and unslipped regions outside a dislocation the Burgers' vector will be an integral number (usually one) of lattice spacings. A dislocation for which this is true is described as *perfect*. There may, however, be *partial* or *imperfect dislocations* in which the original dislocation dissociates into two portions which move together as a unit. The Burgers' vector for each partial dislocation may then be a fraction of a lattice spacing. When this happens the region of crystal between the two partial dislocations will be in the wrong registration with the rest of the crystal. This region is known as a *stacking fault*.

As a simple example, consider Fig. 7·14 in which a section of a close-packed crystal is shown. Suppose that the lower layer comprises the atoms marked A while the next has atoms in the positions marked B and the plane above that reverts to the position A, giving the stacking sequence AB AB AB. Now if slip of the B plane across the A plane occurred each atom would move from one B position to the next. If the move occurred in two stages so that the first was from B to position C and then from C to the next position B the stacking sequence after the first move would become ABACAB. The plane of atoms in the position C would be a stacking fault and the two successive moves would correspond to the passage of two successive partial dislocations.

7·11 Generation of dislocations

The density, ρ, of dislocations in a crystal is usually expressed as length of dislocation line per unit volume, that is, as metres per cubic metre. This is equivalent (although not exactly) to the number of dislocation lines piercing a unit area anywhere in the crystal, so that ρ is usually quoted as lines per unit area and, for convenience, is usually taken as lines per square centimetre. The most perfect bulk single crystal of metal that can be produced will usually have a dislocation density of 10^2 to 10^3 lines per square centimetre while a normal polycrystal has a density in the region 10^7 to 10^8 lines per square centimetre.

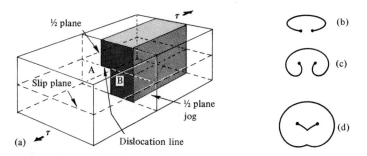

Fig. 7·15 (a) a half-plane jog (b), (c) and (d) operation of a Frank–Read source

These dislocations are built in 'naturally' in the fabrication processes, but when the specimen has undergone severe plastic deformation the density is found to increase to 10^{11} to 10^{12} lines per square centimetre. We must conclude, therefore, that dislocations are somehow generated in the process of deformation. A possible type of 'generator' is called the *Frank–Read* source after its originators.

It is supposed that a dislocation line exists such that each end of it is pinned, that is, unable to move under the particular stress applied. This would be the case, for example, if each end terminated in a half-plane at right-angles to the dislocation half-plane (known as a half-plane jog) as shown in Fig. 7·15(a). Under the influence of the applied stress shown, the line AB would move in the slip plane while there is no resolved component of stress in the plane perpendicular to it which contains the jogs. The result is that the line AB will bow out as shown in Fig. 7·15(b). As the stress increases the line will continue to expand, as in (c), until the two sides meet and join, forming a complete dislocation loop (which represents the boundary between the slipped and unslipped portions of the crystal). The line segment, AB, reforms and the whole process begins again. As the loop expands and passes out of the crystal, a single unit of slip will occur.

7·12 Other crystal defects

Dislocations are, of course, defects in the regular structure of the crystal and much use will be made of the ideas presented in this chapter when dealing with the mechanical behaviour of materials. Reference has been made to vacancies and these will now be considered in more detail, together with other ways in which defects can occur in a crystal structure.

7·13 Vacancies and interstitials

It is important to remember that at all temperatures above absolute zero the atoms in a solid are subject to thermal agitation. This means that they continuously vibrate about their equilibrium positions in the lattice and the amplitude of this vibration increases with increasing temperature. When the vibration becomes strong enough the atoms may break the bonds between them and this corresponds to the solid actually melting and becoming liquid, the corresponding temperature being the melting temperature. Because of this thermal vibration there is always energy available to move an atom from its normal equilibrium position, this coming about by a series of advantageous collisions with neighbours which raise the energy of the atom above its normal equilibrium value. This may be sufficient to move the atom into an *interstitial*† position and it will leave a vacancy on its original site.

Under the influence of thermal energy the interstitial may continue to migrate through the crystal; similarly the vacancy may be filled by an adjacent atom moving into it, leaving the vacancy in a new position so that it also migrates through the crystal. In a normal crystal there will always be a concentration of vacancies and interstitials and these concentrations will be related to the temperature, there being a statistical average concentration for any given temperature. The appropriate statistical calculation is due to Boltzmann and gives the density, n, of vacancies or interstitials as

$$n = N \exp \frac{-W}{kT} \qquad (7 \cdot 11)$$

where N is the total number of atoms per unit volume of the crystal, W is the work done in moving an atom from a lattice site to an interstitial position, T is the absolute temperature, and k is Boltzmann's constant $(1 \cdot 38 \times 10^{-23} \, \text{J}/°\text{K})$. A typical value of W would be about one electron volt $(1 \cdot 6 \times 10^{-19} \, \text{J})$ and for a face-centred metal crystal, such as copper, this gives a value for n/N of about $0 \cdot 001 \%$ at $1,000°\text{K}$. If the metal is quenched (that is, cooled rapidly) from this temperature there will generally not be time for this concentration to drop to that appropriate to room temperature and the metal will have a non-equilibrium concentration of defects. This concentration will gradually diminish over a long period of time by the process of interstitial atoms diffusing until they combine with vacancies, the rate of loss by this mechanism being greater than the rate of generation of new pairs at room temperature.

Vacancies are sometimes referred to as *Schottky defects* and interstitials as *Frenkel defects*. The general term for either type

---

† An interstice is a hole or space between the regularly positioned atoms. An interstitial atom is one occupying an interstice.

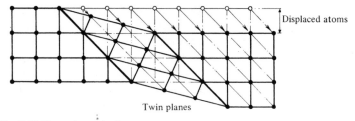

Displaced atoms

Twin planes

Fig. 7·16 Illustrating twinning

is *point defect*. Interstitial atoms which are not of the original parent lattice but are present as a solid solution, as described in the following chapter, are not Frenkel defects.

7·14   Twins

The above dealt with *point* defects; dislocations may be described as *line* defects. We now consider *plane* defects which will be present when the crystal is in some way disordered over a plane. Grain boundaries, described in the previous chapter, may be described as plane defects and so, also, are *twin* boundaries. A crystal is twinned when one portion of the lattice is a mirror image of the other, the 'mirror' being the twinning plane as shown in Fig. 7·16. This may occur naturally in the growth of the crystal or may be the result of mechanical shear stresses. In Fig. 7·16 the successive layers of atoms between the twin planes each have a shear strain which is uniformly distributed amongst the planes in this transition region. This distinguishes it from normal mechanical slip in which there is the same shear displacement of each plane of atoms which forms the block of crystal that has slipped. Furthermore, in slip, the slipped portion has the same orientation as the original crystal whereas in twinning the twinned portion is a mirror image of the original.

The stress to produce twinning tends to be higher than that required for slip but it is less sensitive to temperature and more commonly occurs in metals when they are stressed at low temperatures. This is the case when iron is rapidly loaded at low temperatures when thin, lamellar twin regions appear; these are called *Neumann bands*. Twinning is also an important deformation mechanism in c.p.h. crystals in many of which slip can only occur in the close-packed plane.

*7·15   Dislocation energy

The regularity of a crystal lattice results from each atom taking up a position which minimizes its potential energy; it follows that a dislocation must represent a raising of the potential energies of all the atoms whose positions are affected by its

Fig. 7·17 Model for the calculation of energy of a screw dislocation

presence. Thus an energy may be ascribed to a dislocation which, physically, is the strain energy built into the crystal structure by displacement of the atoms from their regular positions.

We may estimate this energy by considering a screw dislocation in a cylinder of crystal of length, l, as shown in Fig. 7·17. At a radius, r, the deformation in a thin annulus of thickness dr is given by the magnitude, b, of the Burgers' vector so that the shear strain, γ, is $b/2\pi r$ and using Hooke's law for shear, the average stress, τ, will be given by $G \cdot b/2\pi r$ where G is the elastic shear modulus.

It is interesting to note from this that the stress field of a perfect dislocation is inversely proportional to distance from the dislocation core and may, therefore, be described as long range, compared with the interatomic forces. Thus dislocations will interact with each other at quite large distances apart on the atomic scale. In fact at a distance of 10,000 atoms from the core the stress field may still have the value of the yield stress of a soft crystal.

The strain energy dw of a small volume element dv is given by $\frac{1}{2}\tau\gamma\,dv$ so that we have

$$\frac{dw}{dv} = \tfrac{1}{2}\tau\gamma = \frac{1}{2}\frac{Gb}{2\pi r}\cdot\frac{b}{2\pi r}$$

that is

$$dw = \frac{1}{2}G\left(\frac{\mathbf{b}}{2\pi r}\right)^2 \cdot dv$$

The volume of the annular element is $2\pi r l \cdot dr$ so that

$$dw = \frac{G \cdot b^2 l}{4\pi}\cdot\frac{dr}{r} \tag{7·12}$$

The strain energy contained within a cylinder of radius, R, around the dislocation will be obtained by integrating this equation up to the limit, R. However, if the lower limit of integration is taken as zero, the result is that the energy goes to infinity. Such a lower limit is, however, incorrect on two grounds: (1) the integration assumes that an infinitesimal increase in r can be taken at any point, but near $r = 0$ the distances between the atoms and their displacements are finite and dr cannot be taken as vanishingly small compared with distance from the origin of r; and (2) the strains near the dislocation core are exceedingly large and Hooke's law, on which the calculation is based, fails at large strains. Thus the lower limit of integration is taken as some small value, r_0 (which is usually made equal to $\mathbf{b}$). Thus we have

$$E = \frac{Gb^2 l}{4\pi} \int_{r_0}^{R} \frac{dr}{r} + E_0$$

where E_0 allows for the strain energy within the radius 0 to r_0. That is,

$$E = \frac{Gb^2 l}{4\pi} \log\left(\frac{R}{r_0}\right) + E_0 \tag{7·13}$$

It can be shown that E_0 is small compared with the first term in the equation and, since we generally have $R \gg r_0$, the log term varies relatively slowly with R/r_0. As an approximation, it is usual to take $\log(R/r_0) = 4\pi$ so that

$$E \approx G b^2 l \tag{7·14}$$

For a typical metal such as copper, $G \approx 4·55 \times 10\,\mathrm{N/m^2}$ for polycrystalline material and if we take the dislocation length as 1 cm and $b = 10^{-8}$ cm we obtain a strain energy of $2·84 \times 10^8$ eV or in the region of 3 eV per atom in the dislocation line. This is a large energy and a dislocation of length 1 cm would be thermodynamically unstable and therefore would tend to move out of the crystal with the assistance of the thermal vibrations of the atoms.

A second feature of the formula is that the strain energy is proportional to b^2. Thus a large dislocation whose total Burgers' vector was $n\mathbf{b}$ would have an energy density proportional to $n^2\mathbf{b}^2$; this could be lowered by it splitting into n dislocations each with the vector $\mathbf{b}$, when the total energy would be $n\mathbf{b}^2$. Thus large dislocations will, if they can, split up into a number of smaller ones.

A similar calculation for an edge dislocation of length, l, gives the energy as

$$E = \frac{1}{1-v} \cdot \frac{lGb^2}{4\pi} \log\left(\frac{R}{r_0}\right) + E_0 \approx \frac{lGb^2}{1-v} \tag{7·15}$$

where v is Poisson's ratio. This gives energy densities comparable with those calculated for the screw dislocation.

Problems

7·1 A 2,000 kg load is applied to a steel bar of cross-sectional area 6 cm². When the same load is applied to an aluminium bar it is found to give the same elastic strain as the steel. Calculate the cross-sectional area of the aluminium bar. (Young's modulus for steel $= 2·1 \times 10^6$ kgf/cm² and for aluminium $= 0·703 \times 10^6$ kgf/cm².)

7·2 For copper, Young's modulus $= 1·265 \times 10^6$ kgf/cm² and the shear modulus $= 0·352 \times 10^6$ kgf/cm². Calculate the ratio of the tensile strain to the shear strain if the same energy density is produced by a tensile stress alone and a shear stress alone.

7·3 Explain why slip occurs at stresses much lower than those required to pull the atoms of a solid apart. Why does a ductile material elongate plastically under stresses much lower than those required to produce slip?

7·4 Define the Burger's vector for an edge dislocation in a cubic crystal. In a certain crystal the Burger's vector for an edge dislocation is $2·5 \times 10^{-10}$ m. Calculate the force per unit length on the dislocation when a shear stress of 350 kgf/cm² is applied.

7·5 Using the table of physical properties in Appendix 3 together with the Periodic Table of the Elements, plot (a) boiling point against Young's modulus; (b) Young's modulus against valency; (c) density against valency for all the elements which you would regard as metals. What general conclusions can be drawn regarding the relationship between these various factors and the strength of the metallic bonding forces?

8 Mechanical properties

8·1 Introduction

In the previous chapter we considered the mechanical properties of solids from the point of view of atomic mechanisms. We must now consider the bulk macroscopic behaviour of materials under the influence of mechanical forces and attempt to account for them in atomic terms.

In general, solids exhibit two sorts of mechanical behaviour—elastic and plastic. Properties related to elastic behaviour are the elastic moduli defined in the previous chapter and the coefficient of thermal expansion. These are determined by the strength of the interatomic bonds which also determine the brittleness of the material. Plastic behaviour has been described in terms of dislocations and slip and the related bulk properties include ultimate strength, ductility, fatigue and creep behaviour. We begin by discussing the various means of testing and measuring mechanical properties.

8·2 Mechanical testing

8·2·1. *Tensile testing*

As described in the previous chapter, in the tensile test the sample is elongated at a constant rate and the stress and strain are recorded continuously. At the beginning of the test the material extends elastically, the strain being directly proportional to the stress and the specimen returns to its original length on removal of the stress. Beyond the elastic limit the applied stress produces plastic deformation which means that a permanent change in length remains after removal of the stress. As the sample continues to elongate, the measured stress—called the engineering stress—continues to increase as the material work-hardens (that is, strain-hardens) until the ultimate tensile strength is reached. At this point a neck begins to develop in the sample: this is a region where the cross-sectional area begins to decrease and further deformation is concentrated around the neck. After necking has begun the engineering stress recorded on the machine decreases while the elongation continues to increase until the material fractures at the narrowest point of the neck.

Some engineering stress–strain curves typical of different types of material are shown in Fig. 8·1. (For comment on the drop in

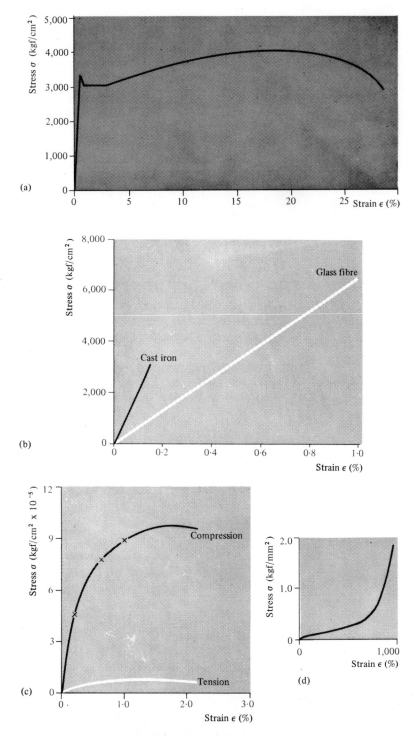

Fig. 8·1 Engineering stress–strain curves for (a) mild steel; (b) brittle solids; (c) concrete in tension and compression; (d) vulcanized rubber

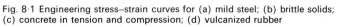

the curve for steel, see section 10·7·1. In ductile materials like steel the engineering stress at fracture is lower than the ultimate tensile strength but in brittle materials such as silicon carbide necking does not occur and the ultimate tensile and fracture strengths are the same. The exceptional behaviour of rubber in giving a very large elastic range is related to its molecular structure. The polyisoprene $(C_5H_8)_n$ molecules, of which it is made, are in the form of long chains (see Chapter 11) having a backbone of carbon atoms bonded directly together with side-links to CH_3 groups and hydrogen atoms. As described in Chapter 11, the chains in such a polymer form an irregular tangle. As they are very flexible, when stress is applied they straighten out and it is only when they have become nearly straight, after large extensions, that the elastic strength of the interatomic bonds comes into play. The elastic behaviour at smaller strains is due to the molecules wriggling about in thermal motion trying to minimize their total energy in opposition to the applied forces.

Returning to the tensile test, since stress is given by the applied load divided by the cross-sectional area, the true stress in the region of a neck will not be the same as the engineering stress which is calculated on the assumption of constant cross-sectional area. To take account of this we define a true stress as the ratio of the load on the sample to the instantaneous minimum cross-sectional area, A_i, supporting the load, P, that is,

$$\sigma_T = \frac{P}{A_i} \tag{8·1}$$

Similarly, the engineering strain is defined as the ratio of the extension to the original length; when the specimen has lengthened considerably this no longer gives the real incremental strain due to an increase in load. We define a true strain as the integral of an incremental change in length, dl, to the instantaneous length, l_i, of the sample: thus

$$\varepsilon_T = \int_{l_0}^{l_i} \frac{dl}{l} \tag{8·2}$$

Thus the true strain of a plastically deformed specimen of initial length, l_0, and instantaneous length, l_i (measured in the absence of stress), is given from equation (8·2) by

$$\varepsilon_T = \log\left(\frac{l_i}{l_0}\right) \tag{8·3}$$

If the measurement is made with the stress applied the contribution of the elastic strain to the total measured strain must be deducted: thus

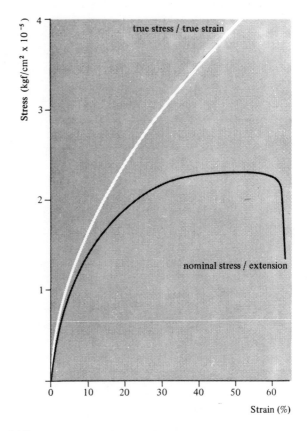

Fig. 8·2 True stress–true strain curve for copper

$$(\varepsilon)_{\text{loaded}} = \log\left(\frac{l_i}{l_0}\right) - \frac{\sigma_T}{E} \tag{8·4}$$

A true stress–true strain curve for copper is shown in Fig. 8·2. Comparison of this with the engineering stress–strain curve indicates that the true stress is not a maximum at the initiation of necking but reaches a maximum only at fracture. It should be noted that the toughness of the material is defined as the area under the curve of Fig. 8·2: hence it is given by $\int\sigma_T d\varepsilon$ in newtons per square metre.

An empirical relationship between true stress and true strain has been found to apply in most cases; this is

$$\sigma_T = C\varepsilon_T^n \tag{8·5}$$

where C is a constant. The exponent, n, is called the strain-hardening exponent and lies usually between 0·1 and 0·3.

Fig. 8·3 Photograph of barrel-shaped distortion produced in a compression test

8·2·2. *Compression test*

The tensile test is almost invariably used for ductile materials but brittle materials are often weak in tension. This is usually due to the presence of microscopic cracks in the material which, when tension is applied, tend to grow in a direction perpendicular to the axis of stress. Such materials can, however, be quite strong in compression; the classic example is concrete for which typical stress–strain curves are given in Fig. 8·1(c).

In compression testing necking does not, of course, occur and the measured curves are very close to true stress–true strain curves. In fact, if the appropriate corrections are made the true curves for both tension and compression should coincide for the same material provided that it is free from internal cracks. Both elastic and plastic behaviour occur in compression, just as in tension, but the compressive test is more difficult to perform accurately for ductile materials which undergo extensive deformation. This is because the surfaces across which the load is applied tend to be constrained by friction between the specimen and the platens of the machine applying the load and the specimen deforms into a barrel shape, as shown in Fig. 8·3.

8·2·3. *Hardness test*

The stress needed to produce plastic flow in brittle materials can be found from indentation hardness tests. The most accurate

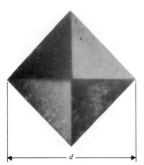

Fig. 8·4 Indention from
a Vickers hardness test

of these is the Vickers test in which a pointed diamond is pushed perpendicularly into the surface by a standard load. The hardness of the surface is expressed as a *Vickers hardness number* (VHN) calculated from an empirical formula. This is related to the size of the indentation of the type shown in Fig. 8·4 made by the diamond which is cut in the form of a square pyramid of apical angle 136°. The length, *d*, of a diagonal between the corners of the rectangular indentation determines the hardness number for a given standard load, *P*, through the formula

$$\text{VHN} = \frac{1.72\,P}{d^2} \tag{8.6}$$

Approximate relationships have been established between the yield stress, *Y*, for plastic flow in compression and the VHN. The forms of these depend on whether Y/E is less or greater than 0·01, where *E* is Young's modulus. As a rough guide, VHN is between two and three times *Y* where *Y* is in kilogrammes per square millimetre for hard materials and $3Y$ for metals. The deformation in a Vickers hardness test is equivalent to a tensile strain of about 8%, so that *Y* refers not to the initial yield stress, but to a value after about 8% strain.

In the similar Brinell hardness test the indenter is a steel or tungsten carbide sphere of 10 mm diameter. This suffers from the disadvantage that the strain varies with the depth of indentation. For consistent results the diameter of the indentation should be between three-tenths and six-tenths the diameter of the ball. Under these conditions fairly soft materials have almost equal Brinell and Vickers hardness numbers.

In the Rockwell test a diamond cone with an apical angle of 120° is used and the depth of the indentation is measured.

Typical Vickers hardness numbers and yield stresses for some hard and soft materials are given in Table 8·1.

Table 8·1   Typical values of VHN and yield stress

Material	VHN (kgf/mm²)	Yield Stress (kgf/mm²)
Diamond	8,400	5,413
Alumina	2,600	1,125
Boron	2,500	1,336
Tungsten carbide	2,100	703
Beryllia	1,300	703
Steel	210	70
Annealed copper	47	No real yield stress
Annealed aluminium	22	No real yield stress
Lead	6·2	No real yield stress

Fig. 8·5 Schematic diagram of an impact testing machine

8·2·4. *Impact test*

Many solids can break in either a ductile manner in which the material stretches before breaking or in a brittle manner in which it does not. The two most important variables involved are the strain rate (the rate at which the material attempts to extend) and the temperature. For example, pitch flows slowly even at room temperature but will break into pieces if hit with a hammer. Glass will flow at elevated temperatures and may be moulded into almost any shape but at room temperature it is brittle. The size and shape of the body are also important; thick pieces of steel may be brittle whereas thin ones are ductile.

These effects are most important in practical designing. The risk of brittle fracture in the vicinity of high stress concentrations, before general yielding sets in, greatly restricts structural engineering design. Many steel ships, especially all-welded ships, have broken in two by brittle fracture round the hull at low temperatures in cold seas.

Tests for brittleness make use of the impact test. In one of these, the Charpy test, a bar of the material to be tested, of length 55 mm and cross section 10 mm × 10 mm, has a notch 2 mm deep cut in one face. The notch angle is 45° and the radius at its peak must be about 0·25 mm. The bar is clamped in a machine shown schematically in Fig. 8·5 and a heavy weight on the end of a pendulum is released from a known height. This strikes the bar on the opposite side to the notch and breaks it. The pendulum swings on and the height to which it rises on the other side is measured. By subtracting the potential energy reached on the overswing from the initial potential energy corresponding to the starting position of the weight the energy absorbed in breaking the bar is determined. The notch brittleness of the specimen is also assessed from the appearance of the fracture, in particular from the proportion of bright crystallite faces on the fractured surface. Typically, toughened steel absorbs in the region of 130 joules of energy in a Charpy test at room temperature. A typical result from notch-tests on mild steel over a range of temperatures is shown in Fig. 8·6. It will be seen that there is a rapid rise in energy absorbed at 280°K as the temperature is raised. This is the ductile-brittle transition; the material is brittle on the low-energy side of the transition. As the temperature rises the material becomes more ductile, so that it flows plastically where hit by the pendulum, and much energy is absorbed. For V-notched specimens the transition temperature is taken as that at which 13–15 joules of energy are absorbed.

8·2·5. *Fatigue test*

It is a familiar fact that if we bend a ductile wire to and fro a number of times it will eventually break. This is a simple example

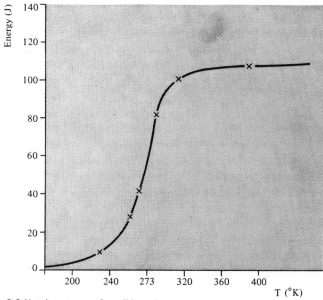

Fig. 8·6 Notch test curve for mild steel

of fatigue failure which will also be familiar to readers of Nevil Shute.† Most failures in service of engineering components subject to alternating stresses are due to metal fatigue. The

Fig. 8·7 Fatigue failure in the wall of a pressure vessel

process is that a small crack forms, usually at a point of stress concentration, and then slowly spreads in a direction across the main tensile stress until a final break occurs suddenly, when the

† Nevil Shute *No Highway*.

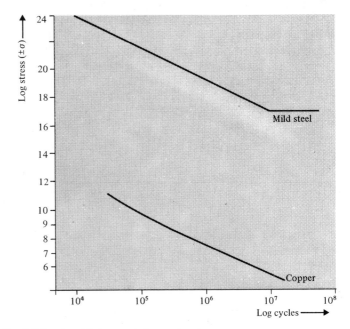

Fig. 8·8 Results of fatigue tests for steel and copper

unbroken cross section has been reduced sufficiently to fail by ductile or brittle fracture.

Fatigue failure is instantly recognizable from the appearance of the fracture, a typical one being shown in Fig. 8·7. Two distinct zones can be seen: the smooth zone is the fatigue crack itself which has been smoothed by the cracked surfaces constantly rubbing together. The rough and crystalline-looking zone is the final fracture. The fatigue crack often shows roughly concentric rings which correspond to successive positions of the crack front.

The fatigue strength of a material is tested by applying a known alternating stress until the sample breaks. The logarithm of the number of cycles, N, completed before failure is plotted against the logarithm of the stress, S, to give a curve of log S against log N, such as those for steel and copper shown in Fig. 8·8.

A simple way to perform this test is to apply a load to the middle of a round bar of the material about one foot long. The load can be applied by hanging a weight from a ball-race placed on the bar which is then rotated by an electric motor on which there is a tachometer to count the number of cycles. This method causes each part of the cross section of the bar (except an axis down the centre) to undergo a stress which alternates between tensile and compressive once for each rotation.

From the curves it will be seen that the allowable stress drops as the number of cycles to be undergone increases. Below a certain value of stress, failure will never occur in steel since the curve apparently becomes horizontal at a stress in the region of 3.5 kgf/cm^2. The point at which the curve flattens out is called the *fatigue limit* and is well below the normal yield stress. For copper, the curve appears smooth throughout and does not tend to any obvious lower limiting value, at least up to a life of the order of 10^9 cycles. Both these properties are currently in dispute; some researchers have observed a discontinuity in the curve for copper; others claim to have found an effective fatigue limit at very low stress levels. The *fatigue strength* is defined as the stress amplitude, S, that produces failure in a given number of cycles, usually 10^7.

Many factors affect the data in fatigue tests and the curves of log S against log N usually show considerable scatter of the experimental points. The most significant factor appears to be surface finish as fatigue cracks are usually initiated on the surface. Thus a polished surface tends to give a higher value for the fatigue strength as also does a chemically or mechanically hardened surface. In addition, fatigue strengths are usually lower in a corrosive environment. A steady tensile stress lowers the fatigue limit while a steady compressive stress increases it.

8·2·6. *Creep tests*

Under the influence of a constant stress, materials continue to deform indefinitely; this phenomenon is called *creep*. Closely related to it is the phenomenon of stress relaxation in which plastic strain slowly replaces elastic strain over a period of time at a constant total strain. Both effects are thermally-activated processes and speed up exponentially with temperature so that

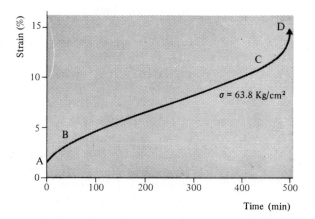

Fig. 8·9 Creep curve of strain against time for lead at room temperature

creep rates more than double for each rise in temperature of 10 degC. This, for example, causes bolts to lose their grip after an extended period at high temperature.

Creep testing can be done on an ordinary tensile testing machine, the strain for a constant applied true stress being plotted as a function of time as in Fig. 8·9. The resulting curve shows four sections. The section OA is the immediate strain, which may be elastic or plastic, and which occurs on application of the load; AB is the transient or primary creep; BC the steady state or secondary creep; and CD is the tertiary creep which ends on fracture at D. At low temperatures and stresses tertiary creep predominates.

For primary and secondary creep the curve of strain against time (ε against t) often fits one of two empirical formulae:

$$\varepsilon = \varepsilon_0 + \kappa t + \alpha \log t$$

or

$$\varepsilon = \varepsilon_0 + \kappa t + \beta t^{2/3} \tag{8.7}$$

In these, ε_0 is the immediate strain (OA) and κt is the steady-state creep, κ being the steady-state creep coefficient: $\alpha \log t$ and $\beta t^{2/3}$ are two forms of transient creep called logarithmic and Andrade creep respectively; α and β are called transient creep coefficients.

The process of creep may be considered as the product of two competing mechanisms: work-hardening, which is discussed in more detail in the next section; and recovery, caused by softening processes such as dislocation climb, thermally activated cross-slip, and vacancy diffusion. We discuss softening processes further in Section 8·4.

8·3 Work-hardening

Mention has been made of the phenomenon of work-hardening (strain-hardening) which is clearly shown in tensile tests, in which the metal becomes stronger as plastic strain proceeds. This is most dramatically shown in single crystals; a single-crystal bar of copper can easily be bent into a ∪-shape by hand but cannot be straightened again. The mechanisms involved can be readily illustrated by reference to a stress–strain diagram for a single crystal (Fig. 8·10). This shows three distinct parts corresponding to different stages of yield. Stage I (referred to as *easy glide*) is entered immediately at the yield point. Here, dislocations glide readily under the applied stress along a single set of parallel slip planes. Strain continues to develop at constant stress and there is little work-hardening because the glide dislocations mostly reach the free surface. After a strain of about 0·01 has been reached Stage II, called linear hardening,

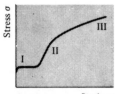

Fig. 8·10 Stress–strain diagram for single crystal of copper

sets in. Here the dislocations are generated and move on intersecting slip planes and become entangled with one another. This may give rise, for example, to jogs and the dislocations become pinned preventing further slip and acting as an obstacle to further dislocation movements in the same planes. Electron micrographs show dense 'forests' of dislocations at this stage. In Stage III (parabolic) hardening, the stress is sufficiently high to activate cross-slip in which screw dislocations can move from one plane to another and then continue to glide giving further strain and a lower rate of strain-hardening.

In polycrystalline materials Stage I hardly occurs at all since the grain boundaries act as effective obstacles to glide. Dislocations pile up against the boundaries until the stress rises to a high enough level to force them over the boundary or to promote a recovery process such as recrystallization, discussed below. Clearly a small grain size will favour work-hardening by introducing more grain boundaries.

High work-hardening strengths can be obtained in metals and alloys which tend to form extended dislocations covering relatively large areas of slip plane. These are less able to slip from one plane to another and bypass obstacles. Bronze, certain brasses, and austentitic steels are in this category. Conversely, crystal structures which favour narrow dislocations, such as copper, gold, and aluminium (f.c.c.) tend to have lower work-hardening strengths at room temperature.

8·4 Recovery processes

The rate of work-hardening usually diminishes with increasing stress because there are recovery, or softening, processes which work in opposition to the work-hardening mechanism.

The disordered structure at grain boundaries can most easily be pictured as a monolayer of a liquid-like structure between adjoining crystals. This can act as a source or sink for lattice vacancies. The emission of vacancies from the boundary is equivalent to moving atoms from inside one crystallite into the liquid layer and then immediately crystallizing them on to the adjoining crystal faces. This process is called vacancy creep and occurs most easily at elevated temperatures; it may also be assisted by stress. The fluidity of the boundary may also allow the sliding of one crystal past another under the influence of shear stress, although, with irregularly shaped grains, this must be accompanied by some plastic deformation of the grains.

The grain boundaries can contribute to recovery in several ways apart from such sliding. It will be remembered that dislocation climb requires the diffusion of vacancies to the end of the dislocation so that, if the boundary emits vacancies into the

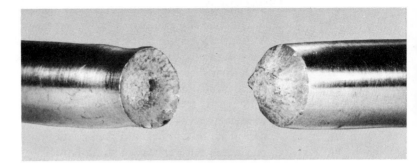

Fig. 8·11 A typical cup and cone ductile fracture resulting from a tensile test

crystallite, climb will be promoted and dislocations can move to new slip planes and continue to glide. The grain boundaries themselves can migrate through the material at elevated temperatures so recrystallizing the material by sweeping up the dislocations in the heavily worked regions ahead of them and leaving relatively undistorted crystal behind them. This is called primary recrystallization and occurs at temperatures near 0·4 to 0·5 T_m, where T_m is the melting temperature of the material in degrees Kelvin. If a material is worked at temperatures below this it is said to be cold-worked. If the working temperature is high enough for recrystallization to occur the material is said to be hot-worked. The material remains soft and the tensile strength usually remains below about $10^{-3}E$, where E is Young's modulus. Pure lead and tin owe their softness to the fact that, for them, room temperature corresponds to a hot-working temperature.

Another factor contributing to recovery is that when dislocations of opposite sign meet each other on the same slip plane they are annihilated. This happens more often when dislocations move from plane to plane, screws by cross-slip, and edges by climb. Thus recovery is assisted by this process as well by the freeing of dislocations. Any process which leads to reduction in the number of dislocations by annihilation or to the rearrangement of dislocations into lower-energy configurations such as sub-grain boundaries assists recovery.

In most metal-working processes there is a limit to the amount of plastic strain to which a part may be subjected without cracking or tearing. For this reason the material is annealed at intervals during the forming process. This consists of heating it above its recrystallization temperature for a period of time τ which allows the formation of new, strain-free grains from which the effects of work-hardening have been eliminated.

Apart from work-hardening, there are several other ways in which the strength of a material may be increased. These include solute and precipitation hardening and fabrication of composite materials. They are considered in Chapter 10.

8·5 Fracture

Two basic types of fracture are recognized: 'brittle' and 'ductile'. Brittle fracture occurs when the material fractures while it is still elastic: it occurs suddenly and the broken bits can be fitted together to form the original shape. Brittle solids usually break at about one-thousandth of Young's modulus: for ordinary glass this is about 7,000 kgf/cm² (10,000 lbf/in²), and for cast iron about 21,000 kgf/cm² (30,000 lbf/in²). Ductile fracture occurs in the plastic region and, at least in tensile tests, is usually preceded by necking. There are three distinct regions in the ductile fracture of polycrystalline materials: (1) necking begins and cavities form in the necked region; (2) the cavities coalesce to form the beginnings of a crack in the centre of the sample; (3) the crack spreads to the surface of the sample in a direction 45° to the tensile axis resulting in a 'cup and cone' type fracture (Fig. 8·11).

The stress at which a material fractures is far lower than the value of the ideal breaking strength calculated from the atomic bond strength. In other words, the fracture strength of real materials is far below the theoretical maximum value for an ideal solid. Why should this be so? Griffith was the first to offer an explanation. He postulated that in a brittle material there are many fine cracks which act to concentrate the stress at their tips. We now deal with how his theory is developed from this idea.

8·6 Griffith's theory

It can be shown that, for a crack of elliptical section (Fig. 8·12) with a stress, σ, applied perpendicular to its long axis of length $2c$, the stress σ_m at its tip is given by

$$\sigma_m = 2\sigma \left(\frac{c}{\rho}\right)^{1/2} \tag{8·8}$$

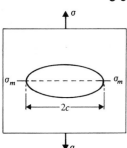

Fig. 8·12 Model for the Griffith crack theory

where ρ is the radius of curvature at the tip. This stress exists within a distance of approximately ρ of the tip and if it exceeds the ideal tensile strength (that is, the strength of the bonds) it may be assumed that the crack will propagate through the material. However, we have also to consider the question of *surface energy*. With every free surface there is associated an energy. We know this because work is needed to create a surface (see Chapter 9) and we have the experimental fact of surface tension. For example, a drop of mercury will always form itself into a sphere unless it is prevented from so doing, because a sphere is the solid shape which has the least surface area for a given volume thus minimizing the surface energy involved. So, as the material separates along a crack, new surfaces are being created and therefore a certain amount of energy must be provided to create them (i.e., some work must be done).

Now before the crack occurs elastic strain energy is stored in the material. This is released when the material relaxes as the crack occurs. Griffith supposed that the crack propagates when the released strain energy is just sufficient to provide the surface energy necessary for the creation of the new surfaces.

The elastic strain energy per unit volume has been given in Eqn (7·8) (Chapter 7) as $\sigma^2/2E$. For purposes of calculation we assume the crack to have unit width in the direction perpendicular to the plane containing the dimension $2c$ (that is, the plane of the paper in Fig. 8·12). Very near to the crack faces the stress falls to zero and very far from the crack it is unchanged, so we assume that roughly a region of radius c around the crack is relieved of its elastic energy. This would, for unit width, give a total elastic energy of

$$\frac{\sigma^2}{2E} \times \text{area} \times \text{width} = \frac{\sigma^2}{2E} \cdot \pi c^2.$$

Properly, the strain field should be integrated from infinity to the surface of the crack, which gives the elastic energy available per unit width as

$$u_E = \frac{\sigma^2 \pi c^2}{E} \tag{8·9}$$

If the surface energy per unit area is γ joules per square metre then the surface energy for a crack of length $2c$ and unit width will be

$$u_s = 4\gamma c \tag{8·10}$$

where we multiply by two because there are two faces. Applying Griffith's criterion means that the change in surface energy with crack length must just equal the change in elastic energy, that is, using Eqn (8·9) and Eqn (8·10)

$$\frac{du_E}{dc} = \frac{du_s}{dc}$$

$$\frac{d}{dc}\left(\frac{\pi c^2 \sigma^2}{E}\right) = \frac{d}{dc}(4\gamma c)$$

that is,

$$\sigma = \sqrt{\left(\frac{2\gamma E}{\pi c}\right)} \div \sqrt{\left(\frac{\gamma E}{c}\right)} \tag{8·11}$$

This result shows that the stress necessary to cause brittle fracture varies inversely as the square root of the crack length. Hence the tensile strength of a completely brittle material is determined by the length of the largest crack existing before loading.

The above analysis applies to a crack in a thin plate under stress but its general features are retained in more complicated analyses and the formula has been confirmed by measuring the strengths of glass containing sharp cracks of known lengths. It is found that the cracks responsible for limiting the usual strength of glass are tiny scratches in the surface. For this reason glass is toughened, for example in car windscreens, by heat treatment that produces a compressive stress in the surface which acts to 'hold the cracks together'.

Problems

8·1 Explain why in a tensile test the level of the true stress–strain curve is higher than that of an engineering stress–strain curve. What would be the relative positions of the two curves in a compression test in which barrel-shaped distortion occurs, and why?

8·2 For what sorts of material is a hardness test most useful? How would a knowledge of the hardness of a material be of help to the machinist, the design engineer, the testing engineer, and the mineralogist?

8·3 In the forging of steel the ingots are heated to high temperatures at intervals during the working process. Discuss the basic mechanisms which are involved in the process.

8·4 How would you distinguish between a ductile and a brittle fracture? Why is the fracture strength of real materials lower than the ideal breaking strength?

8·5 Explain why the strength of a glass plate may be increased by etching off its surface with hydrofluoric acid.

A glass plate has a sharp crack of length 1 micron (10^{-6} m) in its surface. At what stress will it fracture when a tensile force is applied perpendicular to the length of the crack? (Young's modulus $= 0.7 \times 10^6$ kgf/cm^2 and surface energy $= 0.3$ J/m^2.)

9 Thermodynamic properties

9·1 Introduction

In earlier chapters we made frequent mention of 'thermal vibrations' of the atoms in a solid. Our picture is that the atoms oscillate about a rest position with an amplitude that increases with increasing temperature. The question is: how do we know this and, if it is true, what meaning can be attached to it?

The most direct piece of evidence comes from 'Brownian motion' which is named after the botanist Brown. In 1828 he observed that when still air is viewed through a microscope small specks of pollen dust can be seen in continual irregular motion. The cause suggested for this is that the particles are being continually bombarded from random directions by gas atoms which are themselves continually moving about because they have thermal energy. Confirmation of this view lies in the success of the kinetic theory of gases, which is based on the idea of continual motion of the atoms in a gas, in explaining the behaviour of gases.

9·2 Kinetic theory of gases

We picture the gas as a collection of atoms which behave as small elastic spheres—a viewpoint which was justified for solids in Chapter 6. The main idea is that the gas pressure is a result of continual bombardment of the walls of the containing vessel by these atoms. We suppose that each has a velocity, c, and mass, m, so that each carries a momentum, mc, and has kinetic energy $\frac{1}{2}mc^2$. To enable us to calculate the pressure we consider first a single atom contained in a cubical box of side, L. We assume that each collision with the wall is perfectly elastic, that is, the wall behaves like a perfect mirror and the atom bounces off without any loss of energy. If this were not so the energy lost by the atoms when they collide with the wall would cause heating of the wall and this does not happen. If the velocity of an atom is resolved into components, u, v, and w parallel to the sides of the box, then, at each collision, the velocity component perpendicular to the wall is reversed so that u becomes $-u$, and so on. This is illustrated for two dimensions in Fig. 9·1.

If, before a collision, the component of momentum is mu, it will be $-mu$ afterwards so that the momentum of the atom will have changed by $2mu$. By Newton's law of motion, a change

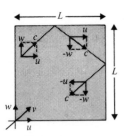

Fig. 9·1 Elastic collisions of a gas atom with the walls of a container

in momentum represents a force. This can be seen from the fact that

$$p = mv \quad \text{and} \quad \frac{dp}{dt} = \frac{m\,dv}{dt} = \text{mass} \times \text{acceleration} = \text{force}$$

so that the force exerted on the wall is given by the rate of change of momentum. Now the atom can only travel a distance L in the direction of u before bouncing off the opposite wall and will return to the first wall after travelling $2L$; thus it will hit the first wall $u/2L$ times per second. As a result we can assume that the change in momentum occurs, on average, in a time $2L/u$ seconds and the rate of change of momentum will be given by

$$\frac{dp}{dt} = \frac{2mu}{2L/u} = \frac{mu^2}{L}$$

Thus the force exerted on the wall is mu^2/L. Since the pressure, P, is the force per unit area of wall due to all the particles with a velocity, u, then

$$P = \frac{\sum mu^2/L}{L^2} = \frac{\sum mu^2}{L^3}$$

that is,

$$P = \frac{m}{V} \sum u^2 \tag{9·1}$$

where $V = L^3 =$ the volume of the vessel. The pressure on the other faces will be found similarly and since the gas is uniform and isotropic these will all be equal.

Thus

$$\sum u^2 = \sum v^2 = \sum w^2 = \frac{1}{3}\sum(u^2 + v^2 + w^2)$$

$$= \frac{1}{3}\sum c^2 \tag{9·2}$$

since

$$c^2 = (u^2 + v^2 + w^2)$$

because u, v, and w are all mutually perpendicular. We may write $\sum c^2$ as $N\bar{c}^2$ where $\bar{c}^2$ is the mean square velocity of an atom and N is the number of atoms in the volume, V. Substituting in Eqn (9·1) we have

$$PV = \frac{1}{3}Nm\bar{c}^2 \tag{9·3}$$

The total kinetic energy, E, of the atoms will be

$$E = \frac{1}{2}Nm\bar{c}^2 \tag{9·4}$$

Having established a relationship between the measurable quantities of pressure and volume and the mean square velocity we can, since we know the mass of the gas molecules, deduce the molecular velocity. This is most easily done by measuring the density of the gas. Since Nm is the total mass of the atoms the density will be given by

$$\rho = \frac{Nm}{V} \tag{9.5}$$

and therefore, combining Eqn (9·3) and Eqn (9·5) the mean square velocity is

$$\bar{c}^2 = 3P/\rho \tag{9.6}$$

From this relationship we may obtain some idea of the mean velocities of the gas atoms. For example at standard temperature and pressure the density of hydrogen is 0·09 kg/m³ (9×10^{-5} g/cm³) and atmospheric pressure is approximately 10^5 N/m² (10^6 dyn/cm²). This gives a root mean square velocity of $1·84 \times 10^3$ metres per second, that is, 1·14 miles per second (4,000 mile/h). A similar calculation for carbon dioxide gives a velocity of 393 m/s at 0°C. Since the mass of the atom is implicitly involved in Eqn (9·6) light particles move faster than heavy ones and this fact is used to separate mixtures of gases of different molecular weight. If the mixture is allowed to escape through a set of small holes the lighter element will escape faster leaving behind a mixture enriched in the heavier element. This is the principle of the gaseous diffusion method used to separate, for example, uranium isotopes for use as nuclear fuels.

It is a familiar fact that when a gas is heated at constant volume its pressure rises in accordance with Boyle's law, which states that the product of pressure and volume is a constant at constant temperature, that is, $PV = RT$, where R is a constant. Thus for constant volume, pressure is proportional to absolute temperature. Now, combining Eqns (9·4) and (9·3) we obtain

$$PV = \frac{2}{3}E \tag{9.7}$$

so that the pressure at constant volume is proportional to kinetic energy and we may conclude that the kinetic energy of gas is proportional to absolute temperature. In fact absolute temperature on the perfect gas scale is *defined* by the relationship

$$\frac{1}{2}m\bar{c}^2 = \frac{3}{2}kT \tag{9.8}$$

where k is Boltzmann's constant and has the value $1·380 \times 10^{-23}$ joules per degree Kelvin. It is a *universal* constant, being the same for all substances.

Substituting from Eqn (9·8) in Eqn (9·3) we obtain

$$PV = NkT \qquad\qquad (9·9)$$

which is the *equation of state* of a perfect gas. If the volume V contains N_0 atoms, where N_0 is Avogadro's number, the equation of state is usually written as $PV = RT$, where R is the gas constant for a kilogramme-molecule. From Eqns (9·4) and (9·8) we also have the total energy of the gas which will be given by $E = \frac{3}{2}NkT$.

9·3   Maxwell–Boltzmann velocity distribution

Our picture of the gas in the foregoing section is that of molecules moving in all possible directions at very high velocities. Naturally the molecules will frequently collide with each other and may gain or lose velocity at each collision although the total energy of the two colliding atoms remains constant. Thus their velocities may vary wildly about an average value but, provided that the gas is in a condition of thermal equilibrium, we can calculate the distribution of particles among the various possible velocities by determining the number having each particular velocity. The detailed calculation is given at the end of the chapter, here we simply quote the results.

Since we wish to express the distribution of the particles among all possible velocities from zero to infinity we can do this by stating the *number* of particles with a *particular velocity*. However, if all velocities are possible we must decide on a minimum separation between the velocities at which we, as it were, count the number of particles. This can be expressed mathematically by letting $f(c)\ \mathrm{d}c$ be the number of particles having velocities between c and $(c+\mathrm{d}c)$ where $\mathrm{d}c$ is a small increment of velocity. The term $f(c)$ is then the number of particles per unit velocity range centred about the value c. It is shown at the end of this chapter that this is given by

$$f(c) = A \exp\left(-\beta c^2\right) \qquad\qquad (9·10)$$

where A and β are constants.

It will be remembered that velocity has a magnitude and a direction and Eqn (9·10) gives the number of particles having a particular velocity in a given direction. We often need to know the total number of particles per unit volume having a velocity in the range c to $(c+\mathrm{d}c)$ irrespective of direction. We define this as $N(c)\ \mathrm{d}c$ and, by a process of integration can show it to be given by

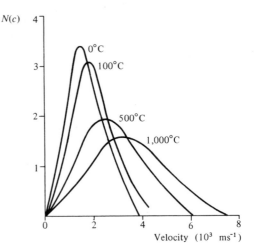

Fig. 9·2 Maxwell–Boltzmann distribution of velocities for hydrogen at different temperatures

$$N(c) = 4\pi c^2 N \left(\frac{m}{2\pi kT}\right)^{3/2} \exp\left(-\frac{mc^2}{2kT}\right) \tag{9·11}$$

where m is the mass of the particle, N is the number of particles per unit volume, T is absolute temperature, and k is Boltzmann's constant. This is the famous *Maxwell–Boltzmann* distribution law and describes, in mathematical terms, how an assembly of fast-moving particles distribute themselves over a range of velocities when they are constantly colliding with each other. We may express the result in the form of graphs of $N(c)$ against velocity, c, as in Fig. 9·2. This shows, for the case of hydrogen, how $N(c)$ varies with velocity at different temperatures and it will be seen that, as temperature rises, the velocities 'spread out' over a wider range. This is to be expected since, if on average they move faster, they will collide more often. By definition of $N(c)$ it follows that $\int_0^\infty N(c)\, dc$ is equal to N, the total number of particles per unit volume. But $\int_0^\infty N(c)\, dc$ is the area under the graph of $N(c)$ against c, so that the area under each of the curves of Fig. 9·2 is the same.

Clearly, since kinetic energy $E = \frac{1}{2}mc^2$, the exponential factor in Eqn (9·11) can be written as $\exp(-E/kT)$. This is a most important fact as, by a general theorem in statistical mechanics, it can be shown that this exponential factor applies to any system of particles in thermal equilibrium. Furthermore the energy, E, in the expression is not limited to kinetic energy only but is the total energy (kinetic plus potential) of the particle. In general we may write for the total number $N(E)$ per unit volume of particles having an energy in the range E to $(E+\delta E)$, that

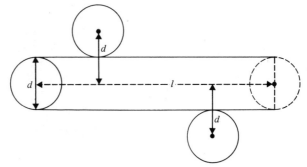

Fig. 9·3 Path of a moving molecule

$$N(E) = A \exp \left(\frac{-E}{kT} \right) \tag{9.12}$$

where A is treated as a constant and depends on the particular system to which the equation is being applied. In most cases A is a slowly-varying function of temperature but its temperature-dependence is ignored because of the extremely rapid variation with temperature of the exponential term.

9·4 Mean free path

In their random motion the molecules of a gas will travel an average distance between successive collisions called the *mean free path*. We may estimate the approximate value of the mean free path by assuming all the molecules to be at rest except one. If all the other molecules are randomly distributed this assumption does not affect the result. The mean free path depends on the diameter, d, of each molecule and the number, N, of them per unit volume. The moving molecule will collide with any other molecule whose centre is within a distance, d, of the path of the centre of the moving molecule as shown in Fig. 9·3. In travelling a distance, l, it will collide with all molecules whose centres lie within a cylinder of length, l, and radius, d, and the number of molecules in such a cylinder will be N times the volume of the cylinder, that is, $N\pi d^2 l$. This will be the number of collisions. Hence the mean free path, λ, will be the length travelled divided by the number of collisions, and is therefore,

$$\lambda = \frac{l}{N\pi d^2 l} = \frac{1}{\pi d^2 N}$$

In fact if the calculation is done rigorously, taking account of the Maxwellian distribution of velocities given in Fig. 9·2, the mean free path is given by

$$\lambda = \frac{1}{\sqrt{2}\,\pi d^2 N} \tag{9.13}$$

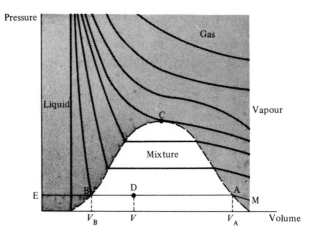

Fig. 9·4 *P–V* diagram for an imperfect gas

The diameter, *d*, is not the actual diameter of the molecule but rather its *effective* diameter for collision. This takes into account various effects such as the compressibility of the particles and the fact that a particle may retain some of its motion after collision. A typical average value for air would be $d \approx 5\text{Å}$. This gives $\lambda \approx 500\text{Å}$ at a pressure of one atmosphere at room temperature and $\lambda \approx 40$ cm at 10^{-4} mm Hg pressure. Combining this with the known root mean square velocity gives the number of collisions per second as being about 10^{10} in air at S.T.P.

9·5 Condensation

So far we have treated only the perfect gas, by assuming Boyle's law ($PV = RT$) to apply. This, in fact, is based on the assumption that the molecules in the gas do not attract each other and that the total volume of the molecules themselves is a negligible fraction of the total volume occupied by the gas. If the gas is compressed so that it occupies a small volume at a high pressure, these assumptions are no longer valid. The actual volume taken to be occupied by the gas must be reduced by an amount corresponding to the volume of the molecules themselves, and the actual measured pressure must be increased by an amount determined by the van der Waals attractive forces between the molecules which tend to pull them together. If isothermals (graphs of pressure against volume for constant temperature) are drawn for a real gas at various temperatures the result is as shown in Fig. 9·4.

At high temperatures a hyperbolic graph is obtained, indicating that the gas obeys Boyle's law. At low temperatures, on the other hand, a horizontal straight portion is found in the curve. As the pressure is raised along the curve MA the volume is reduced until, at the point A, the volume continues to shrink with no further increase in pressure, along the line AB. This occurs when the molecules have been pushed sufficiently close together for the van der Waals attractive forces between them to 'take over' and pull them together to a much denser state. The gas has then *condensed*. The new state is characterized by requiring a large increase in pressure to produce a small change in volume and it does not expand to fill the chamber it occupies; it is, of course, a liquid.

For a particular temperature, the *critical temperature*, T_c, the discontinuity in the isothermal just disappears, as specified by the point C in Fig. 9·4. In the region of C, called the *critical point*, the liquid and its vapour are so similar that they become intermingled, there being no clear interface between them. At other points, such as the point D, we have two distinct *phases*— the gaseous and the liquid—coexisting in dynamical equilibrium with each other. The equilibrium is established such that the rate at which vapour condenses to liquid is equal to the rate at which the liquid evaporates. This equilibrium can be expressed in terms of the isothermal in the following way.

Let the volume of the fluid when entirely in the gas phase at A be V_A and when entirely in the liquid phase at B be V_B. At the point D, let C be the proportion of vapour and $(1-C)$ be the proportion of liquid; then if V is the total volume of the system at the point D

$$V = CV_A + (1-C)V_B$$

that is,

$$C = \frac{V-V_B}{V_A-V_B} \tag{9·14}$$

From the diagram it will be seen that $V-V_B$ is the length DB and V_A-V_B is the length AB. Thus

$$C = \frac{DB}{AB} \quad \text{and} \quad (1-C) = \frac{DA}{AB} \tag{9·15}$$

and the proportions of the two phases can be derived directly from the graph. This is called the *lever rule*.

Clearly, then, the boiling point of a liquid for a given pressure will be the temperature at which the horizontal portions of the isotherms begin at that pressure. For example, if the pressure is that corresponding to the line ABE, then as the temperature is increased from a low value at constant pressure, the volume of the liquid increases along the line EB until the point B at which vapour begins to form and the liquid starts to boil. The value of pressure at this point is called the *vapour pressure* of the liquid and when this is in the region of one atmosphere the boiling point is about $\frac{2}{3}T_c$ (T_c in degrees Kelvin). All substances have roughly similar volumes at the critical point, about 6×10^{-5} m³/mole which corresponds to about 10^{28} molecules per cubic metre.

As in a gas, the molecules in a liquid are still freely in motion with kinetic energies dependent upon the temperature. The van der Waals forces between the molecules are, however, considerable and limit the motion of the molecules preventing the liquid from expanding to fill its container. They also account for the phenomenon of *surface tension*. Molecules in the body of the liquid are subject to equal attractions in all directions from all their neighbours. In the surface, however, there are more neighbours below than above and there is a net attractive force inwards known as the surface tension. If the surface area has to be increased, work must be done against the surface tension force. The product of the force per unit length and the area created is the *surface energy* of this area—a concept which was used in the Griffith crack theory discussed in Chapter 8.

9·6 Thermodynamic equilibrium

So far we have discussed only the kinetic energy of molecules in a gas and have seen that this is the same as the heat energy stored by the gas. It is this kinetic energy which gives rise to the pressure exerted by the gas on the walls of its containing vessel. However it is not the only energy in the gas, as we shall now show. Imagine a gas in a container, one of whose walls is free to move against the pressure exerted by the gas. A suitable container consisting of a cylinder and a piston is shown in Fig. 9·5. Let us increase the energy in the gas by supplying heat to it while maintaining the pressure constant. As the temperature of the gas rises its volume must increase in order to prevent the pressure rising. The piston is therefore pushed up against gravity and work is therefore done on it. This work is equal to the force ($P \times$ area of piston) multiplied by the distance moved. That is

$$\text{work done on piston} = PAd = P\Delta V$$

where ΔV is the change in the volume of the gas.

Now the heat supplied to the gas must both raise its temperature and also do the work, $P\Delta V$. The amount of energy supplied is thus $\Delta E + P\Delta V$, where ΔE is the increase in internal energy. This amount of energy can be obtained from the gas by allowing the reverse process to occur so that the energy is effectively stored in the gas. If now we imagine this reverse process to end

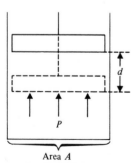

Fig. 9·5 Illustrating work done by gas

in a state of zero volume and zero internal energy, we can see that the total stored energy is just

$$H = E + PV \qquad (9 \cdot 16)$$

where the quantity, H, is the *enthalpy*.

The above description implies that all the energy, H, may be extracted from the gas and that mechanical work and heat energy are freely interchangeable. This cannot be unconditionally true. If it were it would be difficult, for example, to trust hot water to stay where it was put. It might convert some of its heat energy to kinetic or potential energy and move somewhere else of its own volition! Thus we must conclude that heat cannot be converted into work without the accompaniment of some other changes. This is a general statement of the *Second Law of Thermodynamics*.

We can obtain some idea of these 'other changes' by thinking about what happens when a solid melts. We know that latent heat is absorbed and that the temperature does not change during melting. How is this heat stored in the liquid? Since the temperature does not increase the internal energy does not change. Nor does the energy, PV, increase by more than an infinitesimal amount due to the increase in volume. Indeed, when some solids melt, for example ice, a *decrease* in volume occurs. So the latent heat energy is stored in some other way.

We know that the melting of a solid involves an increase in the *disorder* of the molecular arrangement. It is in creating this disorder that the latent heat energy is used up. We must thus equate increasing disorder with an increase in stored energy— energy which is not, however, available for doing work. We call this unavailable energy '*entropy energy*'† and it is conventionally expressed as the product of the temperature, T, and the *entropy*, S, itself. Since the liquid now contains unavailable energy this must be subtracted from its total energy (the enthalpy) if we wish to know the net amount of energy which is available for doing work. Thus we define the *free energy*, G, (often called the Gibbs free energy) by the equation

$$G = E + PV - TS \qquad (9 \cdot 17)$$

The full analytical treatment of this subject is to be found in books on thermodynamics. It is sufficient here to know that the quantity TS represents energy which is 'bound up' in the solid and that the entropy, S, increases as the disorder in the material increases. Like other energies discussed elsewhere in this book, for example in connection with the hydrogen atom in Chapter 3, the Gibbs free energy is a minimum when the system is in

† Entropy means literally 'transformed energy', from the Greek word *trope* meaning transformation.

equilibrium. Since $E = \frac{3}{2}kT$ we can see that, for constant T, P, and V, the free energy, G, will be a minimum when S is a maximum and we conclude that entropy will always tend to a maximum value as the system tends to equilibrium. We can now see why we can trust hot water to stay put since for it to move it must convert some of the free energy into mechanical work. This would reduce the value of G and, at constant temperature, pressure, and volume, would require an increase in the value of S. Since S has its maximum possible value (for those values of T, P, and V) when the water is in equilibrium at its original position this spontaneous work cannot occur.

Statistical mechanics interprets the entropy of a system as a property proportional to its degree of disorder. The condition of perfect order is taken to be one of zero entropy. For example, at the absolute zero of temperature the atoms in a truly perfect crystal would each be stationary in a regular structure and would be perfectly ordered; the entropy of the system would be zero. As the temperature rose the atoms would begin to vibrate with thermal energy and the system would become more disordered, its entropy increasing. If the crystal actually melted, the atoms in the liquid would become completely disordered and this would correspond to a large increase in entropy. As entropy tends to a maximum at equilibrium so we conclude that disorder will also tend to a maximum. This is a natural law and, in part, explains why it is quite difficult to produce single crystals of material.

9·7 Equilibrium between phases

In discussing Fig. 9·4, we saw how different phases of a material may coexist. In general, phases are distinguishable conditions of a material, e.g., vapour, liquid, or solid. However, where different crystallographic forms exist within a solid, these may also be regarded as different phases. The equilibrium phase for a material at a given pressure and temperature will be the one for which the free energy is a minimum, and we can plot a diagram in which we represent the phases as a function of P and T as shown in Fig. 9·6 for water. This is known as a *P–T phase diagram* and the solid lines delineate the pressures and temperatures at which the phases on either side of them are in equilibrium. At the point T it will be seen that all three phases, ice, water, and vapour, are in equilibrium. This is called the *triple point*. Imagine the temperature being steadily reduced at constant pressure along the line ABC, and consider the free energy $G = E + PV - TS$. Along AB the vapour is steadily reducing its internal energy and volume so that the enthalpy is reduced. At the same time the entropy term is falling due to

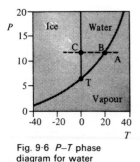

Fig. 9·6 *P–T* phase diagram for water

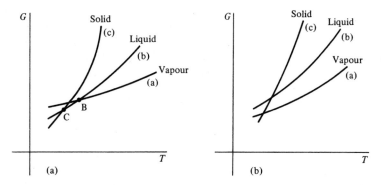

Fig. 9·7 Hypothetical free-energy diagrams for (a) normal vapour–liquid–solid sequence and (b) direct condensation from vapour to solid

reduction in T and to a decrease in S by virtue of decreasing disorder of the molecules. The total free energy of the vapour decreases slowly as the temperature is continually lowered along a curve such as that shown at (a) in Fig. 9·7(a). The free energy of the material, if it could take the form of a liquid or a solid at high temperatures, would also fall for the same reasons. If we suppose their (hypothetical) free-energy curves are as shown at (b) and (c) in Fig. 9·7(a) then, when the temperature has reached the point B, the liquid is the form for which the free energy is lowest and the vapour condenses to liquid form. As the temperature continues to fall the point C is reached and now the free energy is lowest for the solid state so the liquid freezes.

Vapour can, of course, condense directly to crystal, as when hoar frost forms in cold dry weather. Equally, a crystal can sublime directly to vapour as when solid carbon dioxide (dry ice) evaporates in air at room temperature. These two cases correspond to energy curves of the type shown in Fig. 9·7(b) and these will apply to the low pressure and low temperature region in Fig. 9·6.

Mention was made earlier of the existence of different phases in the solid state. Transformations from one to another are known as *allotropic* and the phases themselves are *allotropes*. These occur in many solids such as carbon (already mentioned in Chapter 6), quartz, alumina, tin, and, most important, iron. Since, around atmospheric pressure, the solid phase equilibria in iron are relatively insensitive to pressure, the P–T phase diagram can be simplified to one showing only temperature, as in Fig. 9·8. On cooling, molten iron solidifies into a body-centred cubic structure at 1,539°C. At 1,400°C this undergoes a phase change to a face-centred cubic structure and at 910°C changes back to body-centred cubic which is non-magnetic. At 770°C ferromagnetism appears but is not accompanied by any further change in crystal structure. These phases have been

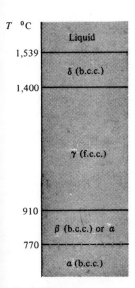

Fig. 9·8 Temperature phase diagram for iron

named α, β, γ, δ from room temperature upwards but the custom nowadays is to label both phases below 910°C as α. The addition of a second component, such as carbon, to the iron modifies the transition temperatures and discussion of this and other alloys is given in Chapter 10.

*9·8   Maxwell–Boltzmann distribution law

We begin by supposing that, in a gas, the atoms behave like perfectly elastic spheres, so that no energy is 'lost' when they collide with each other. Moreover, we shall assume that their motion is entirely random and that the laws of probability can be applied in calculating average velocities and energies.

Let us consider that two particles of equal mass, moving with velocities c_1 and c_2, undergo a collision and then move away with the new velocities c_3 and c_4 respectively as in Fig. 9·9(a). The collision being perfectly elastic, kinetic energy must be conserved so that

$$c_1^2 + c_2^2 = c_3^2 + c_4^2 \tag{9·18}$$

The chances of such a collision occurring in a gas in which only a proportion of the atoms actually have velocities c_1 and c_2 will be proportional to the numbers, $f(c_1)$ and $f(c_2)$, of particles having those velocities. Thus the rate at which such collisions occur will be given by $af(c_1)f(c_2)$ where a is a proportionality factor.

We now imagine the particles to undergo precisely the reverse process. This means that they come in along their 'output' paths with velocities c_3 and c_4 and, after collision, go out along their 'input' paths with velocities c_1 and c_2 as shown in Fig. 9·9(b). Since the gas is in equilibrium, the rate at which this occurs, given by $a'f(c_3)f(c_4)$ where a' is the new proportionality factor, will be precisely the same as the rate at which the first type of collision occurs. Furthermore, since the same paths are traced out, the two collisions are completely equivalent and we must have a' equal to a. Thus

$$f(c_1)f(c_2) = f(c_3)f(c_4)$$

or, taking logs

$$\log f(c_1) + \log f(c_2) = \log f(c_3) + \log f(c_4) \tag{9·19}$$

Comparison of this with Eqn (9·18) immediately suggests a solution of the type

$$\log f(c) \propto c^2 \quad \text{or} \quad f(c) = A \exp(-\beta c^2) \tag{9·20}$$

where A and β are constants.

This demonstration of Eqn (9·20) cannot be considered as

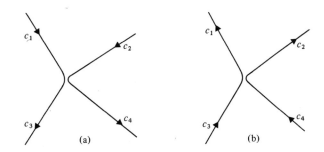

Fig. 9·9 Reversible elastic collisions

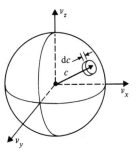

Fig. 9·10 Velocity space

rigorous, but more sophisticated mathematical techniques show it to be the only solution. The negative sign for β is necessary since we require $f(c)$ to diminish as c^2 increases (that is, there are fewer particles at higher velocities).

To obtain an expression for the number of particles having a given speed irrespective of direction we make use of the concept of velocity space. This is simply a three-dimensional space in which the three coordinates are velocity instead of distance as shown in Fig. 9·10. Since the gas is isotropic (that is, has no directional properties) there is, on average, the same number of particles having a velocity, c, *moving in any direction* at any one time. If we draw a vector, $\mathbf{c}$, in the velocity space diagram for every particle with that velocity the vector tips will all lie on a sphere of radius, c. Now the number of particles with a velocity c in a given direction is $f(c)$ so that the total number, $N(c)$, of vectors ending on the sphere is simply $f(c)$ multiplied by the surface area of the sphere. Thus

$$N(c) = 4\pi c^2 f(c) \tag{9·21}$$

The number $N(c)\,dc$ with velocities in the range c to $(c+dc)$ will be the number contained within a spherical shell of radius c and thickness dc. If we add up the volumes of all these shells from zero to infinite radius we will then have covered the total number of particles, N. That is

$$N = \int_0^\infty N(c)\,dc \tag{9·22}$$

Since each particle of velocity, c, has an energy $\tfrac{1}{2}mc^2$ the total energy of the assembly will be given by

$$E = \int_0^\infty \tfrac{1}{2}mc^2 \, N(c)dc \tag{9·23}$$

Using Eqns (9·20) and (9·21) we have

$$N(c) = 4\pi c^2 \, A \exp(-\beta c^2)$$

and substituting in Eqn (9·22)

$$N = \int_0^\infty A4\pi c^2 \exp(-\beta c^2) \, dc \tag{9·24}$$

Similarly Eqn (9·23) becomes

$$E = \int_0^\infty A \tfrac{1}{2}mc^2 \cdot 4\pi c^2 \exp(-\beta c^2) \, dc \tag{9·25}$$

We make use of the standard integrals

$$\int_0^\infty x^2 \exp(-\beta x^2) \, dx = \frac{1}{4\beta}\left(\frac{\pi}{\beta}\right)^{1/2}$$

$$\int_0^\infty x^4 \exp(-\beta x^2) \, dx = \frac{3}{8\beta^2}\left(\frac{\pi}{\beta}\right)^{1/2}$$

With the help of these Eqn (9·24) becomes

$$N = \frac{\pi A}{\beta} \cdot \left(\frac{\pi}{\beta}\right)^{1/2} \tag{9·26}$$

and Eqn (9·25) becomes

$$E = 2\pi \, Am \cdot \frac{3}{8\beta^2}\left(\frac{\pi}{\beta}\right)^{1/2} \tag{9·27}$$

Now using Eqn (9·8) we have $E = \tfrac{3}{2}NkT$ so that

$$\tfrac{3}{2}kT \cdot \frac{\pi A}{\beta}\left(\frac{\pi}{\beta}\right)^{1/2} = 2\pi \, Am \cdot \frac{3}{8\beta^2}\left(\frac{\pi}{\beta}\right)^{1/2}$$

from which

$$\beta = \frac{m}{2kT} \tag{9·28}$$

Substituting Eqn (9·28) in Eqn (9·26) gives

$$A = N\left(\frac{m}{2\pi \, kT}\right)^{3/2} \tag{9·29}$$

so that, finally,

$$N(c) = 4\pi \, c^2 \, N\left(\frac{m}{2\pi \, kT}\right)^{3/2} \exp\left(\frac{-mc^2}{2kT}\right) \tag{9·30}$$

This is the Maxwell–Boltzmann velocity distribution which was quoted in Eqn (9·11).

Problems

9·1 Using Avogadro's number, calculate the number of atoms in 1 g of iodine. At a temperature of 300°C and a pressure of 1 bar, a gramme of iodine gas occupies a volume of 185 cm³; determine the number of atoms in an iodine molecule.

9·2 Let n be the number of gas atoms per unit volume in the earth's atmosphere at a height h above the surface. The pressure, p, at this height is equal to the weight of all the particles above, so that

$$\int_h^\infty n \cdot mg \, dh = p$$

Show, by differentiating this expression with respect to h, that the pressure at any height h is given by

$$p = p_0 \exp \frac{-mgh}{kT}$$

where p_0 is the pressure at the surface.

At what height will the air pressure have decreased by $36\cdot8\%$ at a constant temperature of $300°K$ if the mean mass of the air molecules is 5×10^{-26} kg?

9·3 The most probable velocity of a particle obeying the Maxwell–Boltzmann distribution law is defined as that for which $N(c)$ is a maximum. Show that this velocity is given by

$$\left(\frac{2kT}{m}\right)^{1/2}$$

9·4 Calculate the mean free path for air at a pressure of $0\cdot1$ N/m² and temperature $290°K$, taking the mean radius of an air molecule to be 5 Å.

9·5 Discuss the second law of thermodynamics and the concept of entropy as a measure of disorder.

10 Alloys and ceramics

10·1 Introduction

The most important materials in structural and mechanical engineering are generally alloys or mixtures, that is, combinations of two or more pure materials. The study of what alloys may or may not be formed is thus an important province of materials science and in this chapter we discuss the basic methods involved in such studies. We begin by considering what happens when two pure materials are combined.

10·2 Mixtures and solid solutions

When two liquids, including liquid metals, are mixed together, the resulting mixture will be one of the following three types.

(1) *A solution* in which a homogeneous mixture is formed with the atoms or molecules of one substance randomly dispersed in another. The element in excess is called the solvent and the other the solute—example, water and ethyl alcohol. The mixing of two miscible components into a single homogeneous solution is in all cases an irreversible process.

(2) *A partial solution* in which each liquid is partially soluble in the other so that if there is a small amount of solute and a large amount of solvent a single homogeneous solution is formed. If more solute is added, so that the limit of solubility is reached, two solutions form which, on standing, separate into two layers. Each layer will have one constituent as the solvent with a limited quantity of the other dissolved in it—example, phenol and water.

(3) *A mixture* in which neither component dissolves in the other—example, oil and water (although most oils can dissolve a very small amount of water and vice versa).

In the solid state a mixture of two elements may crystallize in any of the three categories above. In addition, certain compositions of a mixture may behave as crystalline compounds of fixed composition, with a unique crystal structure and sharp melting point. These are known as *intermediate* or *intermetallic compounds*.

In a solid solution the solute atoms are distributed throughout the solvent crystal, the crystal structure commonly being that of

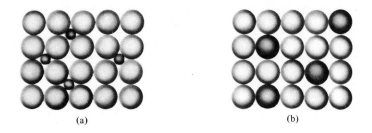

(a)   (b)

Fig. 10·1 (a) Interstitial and (b) substitutional solid solutions

the pure solvent metal. The solute atoms can be accommodated in two different ways. If they occupy interstitial positions, as shown in Fig. 10·1(a), we have an *interstitial solid solution*. Alternatively, if they replace the solvent atoms as shown in Fig. 10·1(b) the resulting combination is called a *substitutional solid solution*. Which of these is formed depends in part upon the relative sizes of the solvent and solute atoms. The situation is similar to the choice of structures in ionic solids discussed in Chapter 6. In general interstitial solid solutions can only form when the solute atom diameter is 0·6 or less of the atomic diameter of the solvent.

The distances between atoms in metals, as between ions in ionic crystals, approximately obey an additive law, each atom or ion being packed in a structure as if it were a sphere of definite size as discussed in Chapter 6. Actually, the radius of the sphere for any atom or ion is not a constant size but varies according to the number of neighbours it has, that is, it depends on the coordination number defined in Chapter 6. There is, for example, a 3% contraction in the radius when passing from 12-fold (close-packed) to 8-fold coordination. However, for purposes of comparison, atomic radii given in books of tables are based on the size for 12-fold coordination. In Table 10·1 these radii are given; in this table the figures marked with an asterisk are elements which form crystals of very low, or no, coordination. In these cases the quoted figure is half the smallest interatomic distance. It is emphasized that Table 10·1 gives the radii of complete (neutral) atoms; where an atom has lost one or more electrons it becomes an ion and, naturally, has a smaller radius. It is *ionic* radii which are given in the table in Chapter 6.

Since the commercially important metals range from cobalt (1·25Å) to magnesium (1·6Å) it will be seen that the atoms which can go into interstitial solutions in these metals must have radii less than 0·75 to 0·96Å. This effectively limits the possibilities to the first five elements, hydrogen to boron. It should be noted that this includes carbon and the interstitial solid solution of carbon in iron is the basis of steel.

Table 10·1   Atomic radii

Element	Radius (Å)	Element	Radius (Å)	Element	Radius (Å)
H	0·46	Ir	1·35	Mg	1·60
O	0·60	V	1·36	Ne	1·60
N	0·71	I	1·36†	Sc	1·60
C	0·77	Zn	1·37	Zr	1·60
B	0·97	Pd	1·37	Sb	1·61
S	1·04†	Re	1·38	Tl	1·71
Cl	1·07†	Pt	1·38	Pb	1·75
P	1·09†	Mo	1·40	He	1·79
Mn	1·12†	W	1·41	Y	1·81
Be	1·13	Al	1·43	Bi	1·82
Se	1·16†	Te	1·43†	Na	1·92
Si	1·17†	Ag	1·44	A	1·92
Br	1·19†	Au	1·44	Ca	1·97
Co	1·25	Ti	1·47	Kr	1·97
Ni	1·25	Nb	1·47	Sr	2·15
As	1·25†	Ta	1·47	Xe	2·18
Cr	1·28	Cd	1·52	Ba	2·24
Fe	1·28	Hg	1·55	K	2·38
Cu	1·28	Li	1·57	Rb	2·51
Ru	1·34	In	1·57	Cs	2·70
Rh	1·35	Sn	1·58	Rare Earths	1·73 to 2·04
Os	1·35	Hf	1·59		

†Estimated from half the interatomic distance in the pure material.

When the atoms are more nearly of the same size a substitutional solid solution is favoured. If the atoms differ in size by more than about 14% the solubility range is likely to be restricted and a partial solution will be formed. For complete solubility the two metals should have the same crystal structure in the pure state and the same valencies. When the valencies differ markedly there will be a tendency, as mentioned in Chapter 6, to form intermediate or intermetallic compounds rather than a solid solution. For example, the atomic sizes of Cu (f.c.c.) and As (rhombohedral) differ by only 2% but the valency of the former is one and of the latter is three. This forms the compound Cu_3As and any surplus of either component forms a separate phase of almost pure material. In the case of silver and copper the valencies and the crystal structures (f.c.c.) are the same and the size difference is only 0·2%. A continuous solid solution can be formed all the way from 100% Cu to 100% Ag.

A solid solution in which at least one of the components is metallic is called an *alloy*.

10·3   Equilibrium phase diagrams

When dealing with alloys of two or more metals the internal equilibrium structure for any composition and temperature can be presented in the form of an *equilibrium phase diagram*. The horizontal direction, or abscissa, is calibrated in composition and gives the percentage of one component in the other while the vertical direction, or ordinate, is calibrated in temperature. Such a diagram is plotted from cooling curves and a particularly simple example is that of copper–nickel which forms a continuous solid solution.

Referring to Fig. 10·2(a), the cooling curve for pure copper contains a flat portion AB during which complete solidification takes place at constant temperature with liberation of latent heat. When 20% nickel is added the flat portion is no longer present but there is a transition region of temperature, $A_1 B_1$, with completely liquid mixture at A_1 going to complete solid at B_1. At temperatures in between there is a mixture of solid and liquid which will remain in equilibrium so long as the temperature is held constant. With 60% nickel similar behaviour occurs over the transition region $A_2 B_2$ but when 100% nickel is reached the transition from liquid to solid, $A_3 B_3$, is again flat, occurring at a fixed temperature. From these results the points A, A_1 A_2 A_3, which correspond to the lowest temperatures at which the solution is entirely liquid, can be plotted as in Fig. 10·2(b). This line is called the *liquidus* curve. Similarly, a curve through the points B defines the highest temperatures at which the solution is a solid and is called the *solidus* curve. The resulting graph is the equilibrium phase diagram and gives the relative amounts of solid and liquid at any composition and temperature.

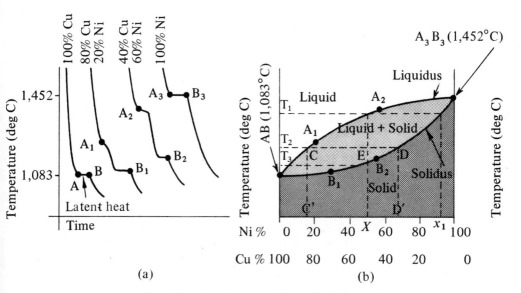

Fig. 10·2 (a) cooling curves; (b) equilibrium phase diagram for copper–nickel alloys

The lever rule, introduced in the last chapter in connection with Fig. 9·4, also applies. Using it, the relative proportions of liquid and solid of different compositions can be found for any temperature. As an example let us consider the composition $X\%$ of copper at a temperature T_2 [Fig. 10·2(b)]. There will be an equilibrium mixture of liquid and solid and from the diagram the equilibrium composition of liquid at temperature T_2 is given by the point C on the liquidus curve so that the liquid in the mixture will have a composition $C'\%$ copper. Similarly the equilibrium composition for the solid phase is given by the point D on the solidus curve corresponding to $D'\%$ copper. The actual proportions of the two phases at the point E is given by the lever rule and will be in the ratio of the lengths CE and ED. In fact:

$$\frac{\text{weight of solid}}{\text{weight of liquid}} = \frac{\text{CE}}{\text{ED}}$$

The rule can be stated in the following way. Let the point representing the composition and temperature be the fulcrum of a horizontal lever. The lengths of the lever arms from the fulcrum to the boundaries of the two-phase field multiplied by the weights of the phases present must balance.

It will be seen from Fig. 10·2(b) that the composition of the solid crystals varies as the temperature falls from T_1 to T_3. The first crystals to form at T_1 will have the composition $x_1\%$ copper and this rises steadily to $X\%$ copper at T_3. Provided that cooling is sufficiently slow for equilibrium to be maintained at every stage, diffusion of atoms from nickel-rich zones to those with a lower nickel content can take place. Thus when the temperature T_3 is reached *all* the crystallites have the same composition: $X\%$ copper. If, however, cooling is too rapid the crystallites may have a very different composition at their centres than at their boundaries, that is, there is a concentration gradient in the crystallites. This may be removed by prolonged annealing at an elevated temperature.

10·4 Partial solid solubility

Most alloys of two metals, although completely soluble in each other in the liquid state, are only partially soluble in the solid state. Such systems form what are known as *eutectics*.

When a liquid containing some dissolved foreign substance freezes, the first crystals to form are usually much more pure than the liquid from which they are formed, as was illustrated in the case of copper and nickel in Section 10·3. This is generally because a liquid with its disorganized atomic structure is able

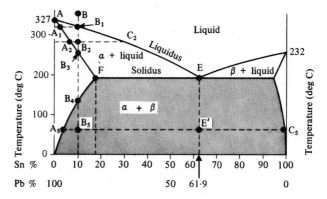

Fig. 10·3 Equilibrium phase diagram for lead–tin alloys

to accommodate foreign atoms, ions, or molecules with less strain than can the solid crystal with its strictly controlled lattice dimensions. The foreign atoms, therefore, try to avoid the crystal so that as crystallization proceeds the liquid becomes richer and richer in the solute substance. This has the effect of altering the composition so that it becomes less and less favourable to crystallization and the freezing point is lowered. At any given temperature, as was seen earlier, there is an equilibrium state between the dilute solid solution and the more concentrated liquid solution.

A possible result of the above process could be a reversal of the roles of solvent and solute. Suppose we have a solution of B in A and, as the temperature is lowered, A starts to crystallize out. The concentration of B in A in the remaining liquid increases and may pass the equal composition point. We now have more B than A in the liquid; B has become the solvent and A the solute. The result as the temperature continues to fall is that there will now be a tendency for the crystals forming to be nearly pure B and the resulting solid will be a mixture of A-like and B-like crystals. This is called a *eutectic structure*. In practice this does not necessarily occur at the point when there are equal quantities of A and B; there is, instead, a *eutectic composition* at which the changeover takes place. (The point E in Fig. 10·3). Thus if the starting liquid has this composition the two phases, A-rich and B-rich, will separate out simultaneously forming eutectic structure at a single temperature called the *eutectic temperature*. A eutectic structure is often in the form of fine plates, needles, or spheres of one phase dispersed throughout a matrix of the other phase.

These ideas are best illustrated by referring to a specific case. In Fig. 10·3 is given the equilibrium phase diagram for the lead–tin system which forms a continuous solution in the liquid state but shows only partial solubilities in the solid state. In the diagram the solid solution of tin in lead is termed the α-phase and that of lead in tin the β-phase.

Let us consider an alloy of 10% Sn and 90% Pb being cooled from 350°C (point B). At B_1 the first crystals of the α-phase form with a composition corresponding to the point A_1 (about 2% Sn). At B_2 the crystals correspond to the composition at A_2, about 5% Sn, while the liquid has a composition corresponding to C^2 at 27% Sn. Using the lever rule,

$$\text{weight of solid} \times (A_2 B_2) = \text{weight of liquid} \times (B_2 C_2)$$

which we will write as

$$S(A_2 B_2) = L(B_2 C_2)$$

then the ratio of solid to liquid is given by

$$\frac{S}{L} = \frac{B_2 C_2}{A_2 B_2}$$

It is more convenient to express the amount of liquid (say) as a percentage of the whole mass $(S+L)$. We write

$$\frac{S}{L} + \frac{L}{L} = \frac{B_2 C_2}{A_2 B_2} + 1$$

that is

$$\frac{S+L}{L} = \frac{B_2 C_2}{A_2 B_2} + \frac{A_2 B_2}{A_2 B_2}$$

Therefore

$$\frac{L}{S+L} = \frac{A_2 B_2}{A_2 B_2 + B_2 C_2}$$

or

$$\frac{L}{S+L} = \frac{A_2 B_2}{A_2 C_2} \tag{10·1}$$

Thus at the point B_2 the percentage liquid present is given by

$$\frac{A_2 B_2 (\%)}{A_2 C_2 (\%)} = \frac{10\% - 5\%}{27\% - 5\%} = \frac{5}{22} \approx 22 \cdot 7\%$$

there being 77·3% of solid.

Proceeding with the cooling, at the point B_3 the entire mass solidifies in the α-phase with 10% Sn. However, as the temperature continues to fall the limit of solid solubility of tin in lead is reached at B_4 and the β-phase begins to form a precipitate in the solid. At B_5 the proportions of α and β phase will be obtained as in Eqn (10·1). That is,

$$\text{amount of } \alpha\text{-phase} = \frac{B_5 C_5}{A_5 C_5} = \frac{100-10}{100-4} = 93\cdot75\%$$

so that there is $6\frac{1}{4}\%$ of precipitated β-phase present. The composition of the α-phase is approximately 4% Sn 96% Pb while that of the β-phase is virtually pure tin.

The eutectic point is shown at E in the diagram and has the composition $61\cdot9\%$ Sn in the liquid phase, forming a mixture of α-phase containing $19\cdot2\%$ Sn and β-phase containing $97\cdot5\%$ Sn. It will be noted that the eutectic composition is the one with the lowest melting point. This alloy is the familiar soft solder used in electrical wiring joints.

We have now considered two different sorts of binary alloy systems in some detail. Some alloys, for example of copper and zinc (brass), show a great many phase changes with temperature and composition and the equilibrium diagram appears very complex indeed. However the principles discussed above can be applied to all such diagrams and the compositions of the various phases can be determined at any temperature.

10·5   Refractory materials

Equilibrium phase diagrams are also useful in studying non-metallic materials. Some of the materials most widely used in high-temperature environments, such as furnaces, are combinations of alumina (Al_2O_3) and silica (SiO_2) the phase diagram for which is given in Fig. 10·4.

These are examples of refractory materials (capable of withstanding high temperature) and alumina–silica compounds occur widely in natural form in the earth's crust. The various phases have therefore been identified by the names given to the natural minerals. Application of the lever rule enables the proportion of phases present at any given temperature to be ascertained from the diagram, from which the following conclusions pertinent to their use can be drawn:

(1) Compositions between approximately 3% and 8% alumina should be avoided as they are close to the region of low eutectic temperature.

(2) The performance of refractories under mechanical load at high temperatures depends on the amount of liquid present at the working temperature, and on its viscosity. For alumina,

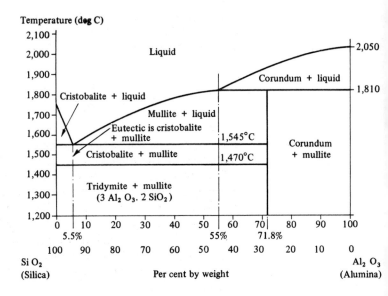

Fig. 10·4 Equilibrium phase diagram for the SiO$_2$–Al$_2$O$_3$ system

percentages between approximately 5·5 and 71·8, the first liquid appears at a temperature of 1,545°C. This liquid will be a smaller proportion of the total amount of refractory material present, the higher is the content of alumina.

(3) For the most severe conditions, where temperature in the region of 1,800°C is encountered, the content of alumina must exceed about 72%.

10·6   Iron–carbon system

Because of its considerable importance in engineering, the iron–carbon system has received much detailed study. The iron–carbon, or more properly the iron–iron carbide phase diagram is given in Fig. 10·5.

The carbon atom is smaller than the iron atom (see Table 10·1) and dissolves interstitially in all three phases, (a, γ and δ), of iron. The solubility in f.c.c. γ-iron is at a maximum near 2·0% at 1,125°C, the solid solution being known as *austenite*. The solubilities in the b.c.c. phases are much smaller, the maxima being 0·1% at 1,492°C in δ-iron and 0·03% at 723°C in a-iron, the latter being called *ferrite*. The alloys involving δ-iron, as will be seen from the phase diagram, cover only a small range of compositions over a narrow range of high temperatures and are not of practical importance. They will not be considered further.

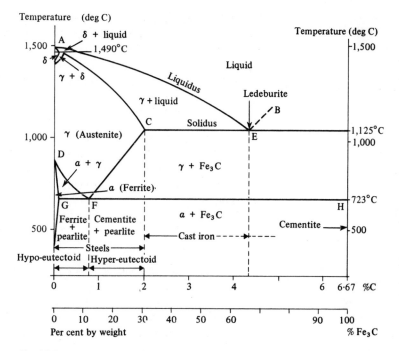

Fig. 10·5 Equilibrium phase diagram for the iron–iron carbide system

   Iron and carbon form an intermediate compound, iron carbide, which contains 6·67% carbon. This has the formula Fe_3C and is called *cementite*. The equilibrium diagram (Fig. 10·5) shows that between iron and cementite there is a eutectic-forming series of alloys with the eutectic point, E, being at 4·3% carbon and 1,125°C. The eutectic is between cementite and austenite and is given the name *ledeburite*. The important engineering materials are restricted to this range of alloys. *Pig iron* is essentially ledeburite. Any alloys with less than 0·03% carbon (that is, entirely ferrite at 723°C) are known as *pure irons*. Alloys in the range 0·03% to 1·7% carbon are called *steels*, while those in the range 1·7% to 6·67% carbon are called *cast irons*.

   Referring to the diagram it will be seen that the lower boundary of the γ (austenite) phase, DFC, is formed by two lines. One is the ferrite separation line, DF, and the other is the cementite separation line, CF. The point, F, at which these intersect represents a second eutectic composition at an equilibrium temperature of 723°C. The precise carbon content of the eutectic is not certain but the figure of 0·89% is generally adopted. The eutectoid forms by simultaneous precipitation of ferrite and cementite and is known as *pearlite*, which often has a laminated form consisting of alternate thin plates of cementite and ferrite.

Steels are often referred to as *hypo-eutectoid* or *hyper-eutectoid* according to whether their carbon content is less than or more than that of the eutectic composition. The proportions of ferrite and cementite in pearlite can be obtained by applying the lever rule along the line GFH in the diagram, that is

weight of ferrite × GF = weight of cementite × FH

This gives the result 13·3% cementite and 86·7% ferrite.

Much further information regarding microstructure, mechanical properties, heat treatment, and so on of steels and cast irons is available. For this students are referred to textbooks on metallurgy.

10·7   Properties of alloys

Alloying has a marked effect on the mechanical properties of alloys. Eutectoids, as might be expected from their structure, have a hardness which changes in a linear fashion with the relative proportions of its two components. The various ways in which the hardness of a metal is increased by alloying are now discussed.

10·7·1.   *Solute hardening*

A common way of increasing the hardness and yield strength of a metal is by dissolving a second metal in it to form a solid solution. If the atoms of the solute metal are bigger or smaller than those of the original material they will set up local strains around them. These strain fields act as obstacles to dislocation motion and hence raise the strength of the material. Random solid solutions provide only mild hardening, however, because on average a glide dislocation is pulled equally to and fro by the solute atoms on either side of it. Non-random solutions may be formed in which the solute atoms enter regular preferred sites to reduce the elastic energy or to increase the number of strong atomic bonds. In this case the movement of dislocations disturbs or destroys this special distribution and this requires extra work, that is, extra stress, so that the material is hardened more than in the case of the random solution. These non-random arrangements may be short range order (or clustering) in which the solute atoms gather together in small groups or they may be long range order in which the solute atoms arrange themselves in a regular pattern throughout the crystal. Since these distributions are produced by diffusion processes in the solid solution, the hardening due to them gradually develops as the alloy is rested or aged at a suitable temperature.

Another type of short range order which can develop in

some materials is segregation at dislocations. This happens in mild steel and many of its unique properties are due to interstitial carbon and nitrogen atoms migrating to dislocations because the interstitial sites there have more room for them. This segregation has the effect of pinning the dislocations so that on the initial application of stress, the strain rises steeply to an upper yield point. At this stress the dislocations break free from the primary clusters and a yield drop occurs, due to dislocation glide, to a lower yield point where yield elongation begins. This is shown in Fig. 8·1(a) in Chapter 8.

Perhaps the most important example of a solute-hardened material is martensite, the hard constituent of quench-hardened carbon steel. The carbon is first taken into solid solution, up to four atomic per cent in a typical steel, by heating to around 900°C where it is austentitic. It is then held in solution by quenching to room temperature. The steel transforms to the b.c.c. a-structure under this treatment, but, due to the rapidity of the quenching, the carbon is forced to remain in supersaturated interstitial solution and distorts the b.c.c. structure to a body-centred tetragonal (Fig 6·14) which is martensite. The carbon atoms are ordered and produce the resultant hardening.

10·7·2. *Precipitation hardening*

If resistance to dislocation motion were all that was required of a strong material we could use intrinsically hard solids like silicon carbide for engineering applications. Such materials are, however, brittle and it is more often a combination of ductility and strength that is required. The ductility not only makes it possible to work the material to a required shape but gives a much greater fracture strength (as opposed to yield strength) than does a brittle material. The best combination can be obtained from an intrinsically soft metal that is strengthened by finely dispersed particles of hard substances. Since these particles are usually produced by precipitation such materials are said to be precipitation hardened.

In certain alloy systems this process can be more effective than hardening by reduction of grain size, cold work or solid solution hardening. In aluminium alloys, for example, the first stage of hardening is to quench them from 500°C so that copper and other elements are retained in supersaturated solid solution. Subsequent heat treatment ('aging') at 150°C causes the copper atoms first to cluster together in the aluminium grains and then to form precipitate particles of an intermetallic compound. Tensile strengths which, in normal homogeneous metals and alloys, are in the region of thousandths of the value of Young's modulus can be increased up to about one-hundredth of Young's modulus by these precipitate particles.

The tempering of steel by aging it between 100°C and 300°C causes carbon to precipitate as fine iron carbide particles, thereby reducing brittleness without much reduction in hardness.

If the precipitate particles are dispersed too finely the glide dislocations can cut right through them and hardness is not much increased. If, on the other hand, aging is carried too far so that the particles are relatively large and widely-spaced the dislocations can pass between them. Thus there is always an optimum time and temperature for the heat treatment to produce the maximum increase in hardness.

10·7·3. *Composite materials*

In the study of precipitation-hardened materials it has been found that the particles are subject to large elastic strains when the material is stressed. Thus the particles contribute to the load-bearing capacity of the material as well as acting as obstacles to dislocation movement. This suggests that if a very large volume fraction of strong particles were included great strength could be produced by use of the load-bearing properties of the particles. The best way of achieving this is to make the particles in the form of long fibres. Under stress the matrix material in which they are embedded may flow but, in so doing, will produce a frictional force on the fibre which, if the fibre is long enough, will ultimately lead to its fracture. Thus the fibres contribute fully to the strength of the composite material. The preferred microstructure should obviously be such that the matrix material flows easily in a direction parallel to the fibres and that all the fibres be oriented in this direction.

The breaking strength of strong solids is, as was seen in Chapter 8, sensitive to the presence of cracks. If the material is in the form of a bundle of fibres then the only cracks that matter must necessarily be short—across the width of the fibre—and are limited to the fibres in which they exist. A solid made up from such fibres embedded in a suitable matrix may therefore be expected to have a greater fracture strength than would a solid lump of the strong material of the same total volume.

The fibres must, somehow, be joined together: hemp fibres can be twisted together to form long strands which are then coiled together to form a rope. Glass fibres can be produced but their strength is very dependent upon surface damage and they cannot therefore be twisted into a rope form. However, if a suitable embedding matrix such as polyester resin is used it can serve to join the fibres together and at the same time protect their surfaces. This is the basis of the very strong and light glass fibre materials now widely used.

By suitable choice of materials strong solids can be made with fibres of extremely hard material, such as silicon carbide, embedded in a metallic matrix. The matrix must fulfil three functions: it must protect the fibre surfaces from damage due to abrasion, it must separate the individual fibres so that cracks cannot run from one to another, and it must be able to bind to the fibre surface so that load can be transferred to the fibre. Research has shown that for best results the fibres themselves must exceed a certain critical length, l_c, and have a length-to-diameter ratio in excess of about $5l_c/d$, where d is the diameter.

Development of these materials is one of the most active areas of materials research at the time of writing. Strengths up to half the maximum tensile strength of the fibres themselves are attainable. For example, whiskers of silicon carbide (SiC) have a maximum tensile strength of $2{\cdot}1 \times 10^5$ kgf/cm^2 (3×10^6 lbf/in^2) at room temperature so that a composite should give a strength of 10^5 kgf/cm^2 ($1{\cdot}5 \times 10^6$ lbf/in^2). This can be compared with the strongest steel wire available which has a strength of $0{\cdot}42 \times 10^5$ kgf/cm^2 ($0{\cdot}2 \times 10^6$ lbf/in^2).

10·8   Alloy density and conductivity

The density of an alloy is, of course, different from that of its constituents taken separately. This is not only due to the differing masses of the two atoms but is also affected by the manner in which they are packed together in the alloy crystal, that is, on the crystal structure. In most continuous solid solutions between two elements of similar crystal structure the lattice constant varies linearly with composition from the value for one pure material to that for the other. This is *Vegard's law*.

For interstitial solid solutions the density of the material will increase by virtue of the additional atoms in interstitial positions. If A is the interstitial solute and B the solvent material, the density will be given by

$$\rho_i = (n_A A + n_B B)\,\frac{m}{V} \tag{10·2}$$

where n_A is the average number of atoms of atomic weight A in a unit cell and n_B is the number of atoms of atomic weight B in a unit cell and is always a constant independent of the value of n_A. The symbol V is the volume of a unit cell and m is the mass of the hydrogen atom.

For a substitutional solid solution the solute species A actually replaces the solvent species B and the density may increase or decrease, depending on their relative atomic weights.

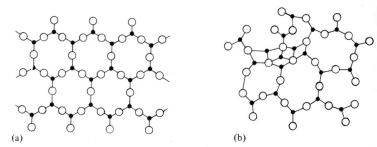

Fig. 10·6 (a) Crystalline lattice structure of silica; (b) network structure of glassy silica. Circles indicate oxygen atoms, black dots indicate silicon atoms.

The density will be given by

$$\rho_s = nM \frac{m}{V} \tag{10·3}$$

where $M = CA + (1-C)B$ where C is the fractional concentration of A in B and n is the number of atoms in a unit cell. Equations (9·21) and (9·22) apply whether the alloy forms a eutectic or not.

The thermal and electrical conductivities of alloys also differ from those of the constituent elements. The electrical conductivity is invariably lowered by the addition of a solute element and in alloys showing complete solid solubility it passes through a minimum and then rises to the value appropriate to the pure solute material. This is explained in Chapter 9. When a eutectic is formed the conductivity is an average of the conductivities of the two components in proportion to their concentrations. Thermal conductivity follows much the same pattern, and the specific heat increases as the thermal conductivity decreases.

10·9 Ceramics and glasses

The word ceramic derives from the Greek 'keramos' which was the name for potter's earth or clay. In modern usage the word has come to be applied to a wide range of non-metallic compounds, usually of a hard brittle nature and generally being in the form of amorphous (non-crystalline) or glassy solids. Such materials as silica (SiO_2) and alumina (Al_2O_3) are typical. The atomic bonding in these materials is of a mixed ionic and covalent character and while they can be made in single-crystal forms (these include quartz in the case of SiO_2 and sapphire in the case of Al_2O_3) their more common structures are amorphous or glassy.

Silica can crystallize in a number of forms all of which can be regarded as a network of oxygen ions, following the cubic or hexagonal type of lattice, with silicon ions in the tetrahedral spaces between them. This is shown schematically in two dimensions in Fig. 10·6(a). If the crystalline form of silica is melted and then rapidly cooled it is unable to attain the long range order of the single-crystal state. Instead it forms a short range ordered network structure as shown in Fig. 10·6(b). This structure is not an equilibrium one and will very slowly tend to change to a crystalline form—a process often aided by the application of stress. The network structure is, in effect, a supercooled liquid and any material forming it is called a glass. It is characterized by having no definable melting temperature since, as the temperature is raised, it simply softens progressively until it becomes an easily flowing liquid. A glass is called a *vitreous* solid and if the material returns to the crystalline state it is said to have been *devitrified*. One of the important features of the glassy structure is that it is a very open network and can easily accommodate many impurities. Silica is the most common compound on the earth's surface and the majority of rocks are basically silicates. The silicon atoms can also readily be replaced by sodium, potassium, calcium, or lead which, in effect, form substitutional solid solutions in small concentrations. They affect the properties of the glass and, because of their different sizes, tend to prevent crystallization taking place. Ordinary window glass contains sodium together with lead or calcium and these impurities appear to assist its formation in sheet form. 'Pyrex' glasses are harder as a result of the incorporation of potassium.

When one-quarter of the silicon ions are replaced by aluminium and potassium ions the material belongs to the *feldspar* class of minerals. Since aluminium is trivalent and potassium monovalent one of each forms a pair which, when it forms bonds, is equivalent to one quadrivalent silicon atom. Weathering of feldspar produces kaolin (potter's earth), the original 'keramos', which is pure hydrated aluminium silicate. The clay minerals owe their plasticity to their water of hydration. Once the clay loses its water, as it does when fired at high temperature, it becomes an amorphous solid. The water cannot easily be reintroduced and so fired pottery remains fairly strong.

The mode of deformation of ceramic materials is highly dependent on structure. The lack of any long range order makes the motion of dislocations impossible. However, these materials can deform by viscous flow processes under proper conditions of stress and temperature. A fluid flows viscously, that is, it cannot support an applied shear stress statically but deforms continuously. The resulting shear strain is then a function of shear stress and time. For an ideal fluid the shear strain, γ, is related to the shear stress, τ, by

$$\tau = \eta \frac{\partial \gamma}{\partial t} \qquad\qquad (10\cdot4)$$

where η is a constant called the coefficient of viscosity, usually measured in poises. A fluid has a viscosity of one poise when a stress of one dyne per square centimetre produces a shear rate of one reciprocal second. In MKS units one poise has the value 0.1 kg/cm s or 0.1 N s/m^2. Typical values near room temperature in poises are: air, 1.8×10^{-4}; water, 10^{-2}; treacle, 10^3; pitch, 10^{10}. In molten glass its value lies between 50 and 100. Blowing, drawing, and rolling can be carried out at a lower temperature when the viscosity is between 10^4 and 10^8 poises, while at room temperature glass has a viscosity of 10^{20} poises.

At low temperatures, where viscous flow is so slow as to be negligible and plastic flow is impossible, glasses and indeed all ceramics behave elastically. They are extremely notch-sensitive at these temperatures and, as was mentioned in Chapter 8, internal and surface microcracks greatly lower their tensile strengths although their compressive strengths may be quite high. They also have a much higher resistance to creep than metals of melting point similar to that of the crystalline ceramic.

One of the chief uses of these non-metallic materials is as insulators, both thermal and electrical. Because all the valency electrons are occupied in the covalent–ionic bonds none is free to conduct electricity or to transfer thermal energy. As a result ceramics are generally good insulators.

10·10   Cement and concrete

Perhaps the most extensively used engineering material is concrete. This is made by mixing mineral lumps (stones), called *aggregate* with a cement paste which binds them together. Cement is a mixture of materials which sets and bonds the aggregate together through the action of its water of hydration. The basic operative materials are the calcium silicates $2CaO.SiO_2$ and $3CaO.SiO_2$ which, when hydrated, form a material called *tobermorite* gel. This consists of a distorted crystalline layered structure with water of hydration dissolved between adjacent layers and has the formula $2CaO.SiO_2.x(H_2O)$, where x may vary considerably depending on the ratio of water to cement.

The most commonly used formulation is that of Portland cement, the constitution of which is given together with the symbols used for the constituents in Table 10·2.

On adding water there is an initial fast reaction

$$C_3A + 6H_2O \rightarrow C_3A.6H_2O$$

which evolves considerable heat of hydration. The resulting hydrate crystals coat the aggregate particles and seal them off so that they do not absorb water. The gypsum tends to slow this reaction so that heat is not evolved too rapidly. The principal subsequent hydration reactions are

$$C_2S + xH_2O \rightarrow C_2S.x(H_2O)$$

and

$$C_3S + xH_2O \rightarrow C_2S.x(H_2O) + Ca(OH)_2.$$

Table 10·2   The constitution of Portland cement

Constituent	Symbol	Weight %
Dicalcium silicate (2CaO.SiO$_2$)	C$_2$S	28
Tricalcium silicate (3CaO.SiO$_2$)	C$_3$S	46
Tricalcium aluminate (3CaO.Al$_2$O$_3$)	C$_3$A	11
Tetracalcium alumino ferrite (4CaO.Al$_2$O$_3$.Fe$_2$O$_3$)	C$_4$AF	8
Gypsum (CaSO$_4$)	—	3
Magnesia (MgO)	—	3
Calcium oxide (CaO)	C	0·5
Sodium oxide (Na$_2$O)	—	0·5
Potassium oxide (K$_2$O)		

The C$_3$S takes about 30 days to reach 70% of its ultimate strength, while the C$_2$S only reaches about two-thirds of its final strength in six months at normal temperatures. Thus concrete goes on hardening for years.

In order to achieve high strength it is important to use the correct ratio of water to cement. If there is too little water, entrapped air gives a porous and therefore weaker structure. Too much water has a similar effect since evaporation of the free water also leaves pores. In modern techniques ultrasonic vibration is used to compact the cement and get the air out of it and this gives about a 15% improvement in compressive strength over that of hand-mixed cement.

Problems

10·1 Using the Sn–Pb equilibrium diagram of Fig. 10·3 and applying the lever rule, answer the following questions:

(a) What are the compositions of the phases at the eutectic composition?
(b) For an alloy containing 65% by weight of Sn, what fraction exists as the β phase at 220° C?
(c) What fraction of the alloy is liquid just above the eutectic temperature?
(d) Determine the fractions of α and β phase just below the eutectic temperature.
(e) What fraction of the total weight of the alloy will have the eutectic structure, just below the eutectic temperature?

10·2 Using the SiO_2–Al_2O_3 phase diagram of Fig. 10·4 answer the following:

(a) What fraction of the total ceramic is liquid for a 30% Al_2O_3 composition at 1,545°C?

(b) What are the compositions of the phases involved in the eutectic reaction?

(c) Describe the changes that take place as a 60% Al_2O_3 composition cools from 2,000°C to room temperature under equilibrium conditions.

10·3 Using the iron–carbon diagram of Fig. 10·5 answer the following questions:

(a) A 0·4%C steel is cooled from 1,600°C to room temperature under equilibrium conditions. State the sequence of phases and the temperatures at which they occur.

(b) Determine the relative weights and compositions of the phases present at 800°C.

(c) What are the weights of ferrite and cementite present in pearlite?

(d) What is the composition of ledeburite?

10·4 Discuss the relative merits of solute- and precipitation-hardened and composite materials.

10·5 Discuss, from the atomic point of view, the factors that determine the mechanical properties of ceramics.

11 Organic polymers

11.1 Long chain molecules

Today we are familiar with a wide range of materials—plastics, rubbers, resins, and fibres—all of which are organic polymers. Such materials are of considerable technological importance and the annual production of polymeric materials is growing at an enormous rate.

Because all their properties are so very different from those of the inorganic solids we have mentioned hitherto it is necessary to discuss them in some detail.

Polymers are molecular materials, that is to say, the constituent atoms combine together in molecules which are then held together by secondary bonds such as the van der Waals or hydrogen bonds. The unique characteristic of a polymer is that each molecule is in the form of an extremely long chain. This is best appreciated by considering polyethylene ('polythene') which is one of the simplest polymers and whose chemical structure is indicated in Fig. 11·1.

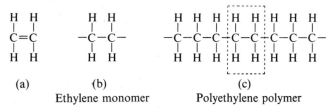

(a)   (b)   (c)

Ethylene monomer   Polyethylene polymer

Fig. 11·1 (a) The monomer ethylene; (b) ethylene monomer with the double bond broken; (c) the polymer polyethylene

A polyethylene molecule is built up by joining together many molecules of the monomer ethylene (C_2H_4). In the isolated molecule of ethylene the two carbon atoms are held together by two covalent bonds each containing two electrons and indicated by a single dash in Fig. 11·1. We saw in Chapter 5 that carbon bonds are very directional and normally prefer to point towards the corners of an imaginary tetrahedron (see, for example, Fig. 5·12) so that the double bond in ethylene is strained. It is relatively easy, then, to break one of the two bonds and create two points at which new bonds may be formed [Fig. 11·1(b)]. In this way two or more ethylene units may be joined.

If the process is continued larger and larger molecules may be built up. The number of units in the molecule is known as the *degree of polymerization,* or D.P. At D.P. values of about 10 to 20 the substance so formed is a light oil—a paraffin if formed from ethylene. As the D.P. increases the substance becomes greasy, then waxy, and finally at a value of D.P. of about 1,000 the substance becomes a solid and is then a true polymer. Naturally the D.P. is almost unlimited—it may increase to around 100,000 or so. The effect of molecular size on the properties of a polymer will be mentioned later in the chapter.

The molecular weight is another measure of chain length and is equal to the D.P. multiplied by the molecular weight of the monomer. Thus polyethylene with a D.P. of 10,000 has a molecular weight of 280,000.

If we begin with another monomer a different polymer is created. For example the monomer vinyl chloride (C_2H_3Cl) can be polymerized to form polyvinyl chloride (p.v.c.) (Fig. 11·2).

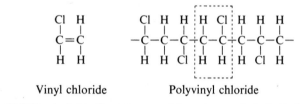

Vinyl chloride   Polyvinyl chloride

Fig. 11·2 The vinyl chloride monomer and the polymer polyvinyl chloride

The two examples given are called linear polymers and we see that they consist of a backbone of carbon atoms[†] and a number of side groups which differ from polymer to polymer. An alternative way of describing a linear polymer is shown in Fig. 11·3(a) in which it is pictured as a covalently bonded chain of monomer units.

In Fig. 11·3(b) we show how, by joining together different monomers in the same chain, it is possible to form a *copolymer.*

------—M—M—M—M—M—M—------
A linear polymer   M = monomer unit, e.g. ethylene
(a)
—L—M—L—M—L—M—L—M—
Regular copolymer   L,M = monomer units
—L—M—L—L—M—M—M—L—M—M—L—L—L—
Random Copolymer
—L—M—M—M—M—M—M—M—M—L—L—L—L—L—L—L—
Block copolymer
(b)

Fig. 11·3 Copolymers may be made with different structures

Different kinds of copolymer may be made by changing the sequence in which the different monomer units appear in the chain.

† The backbone is not invariably of carbon, as we shall see later.

Another kind of polymer may be made by removing a side group and replacing it with a chain. This forms a *branched* polymer and Fig. 11·4(a) shows branched polyethylene. If many such branches are formed a network structure results, as indicated in Fig. 11·4(b), in which the long chains are connected together, often by relatively short *cross-links*.

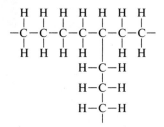

(a) Branched polyethylene

(b) A network polymer (schematic only)

Fig. 11·4 Branching in polyethylene, and a schematic representation of a network polymer

11·2   Thermosetting and thermoplastic polymers

It is convenient at this point to divide polymers into two groups which differ in their structure, their properties, and in the chemical processes used in their manufacture.

A polymer which forms a network, with cross-links between chains is a *thermosetting* polymer. As the name suggests, such a polymer becomes set into a given network when it is manufactured. An item manufactured from a thermosetting plastic cannot subsequently be remoulded to a new shape. If the temperature is raised to the point where the cross-links are broken, irreversible chemical processes also occur which destroy the useful properties of the plastic. This is called *degradation*. At normal temperatures the cross-links make the solid quite rigid.

On the other hand a *thermoplastic* material may be readily moulded or extruded because of the absence of cross-links. As described in the next section, a thermoplastic becomes quite plastic if the temperature is raised and can be moulded into

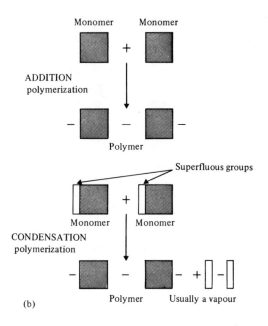

Fig. 11·5 Polymerization processes

shape even at temperatures below the melting point. On the debit side, the mechanical properties of these plastics are rather temperature sensitive, unlike those of a thermosetting material.

The distinction between thermoplastic and thermosetting polymers is also important to the manufacturing chemist since they are prepared by quite different chemical processes.

To prepare a thermoplastic, one begins with the monomer or monomers which are to form the repeating units in the polymer. Under suitable conditions of temperature and pressure and in the presence of a catalyst called an *initiator* the molecular chains grow by the addition of monomer molecules one by one to the ends of the chains. Branching can occur, but cross-links are nearly absent. This process is called *addition polymerization* (Fig. 11·5).

Thermosetting polymers are prepared by the process of *condensation polymerization* (Fig. 11·5). In this case the starting monomers are not identical to those of which the chains are to be composed but contain some superfluous groups of atoms which must be ejected when the unit is added to the end of the chain. If there are enough groups of such superfluous atoms on each monomer molecule some of them may be temporarily retained on the side of the chain as it grows. This promotes easy branching, and leads rather rapidly to a highly cross-linked structure.

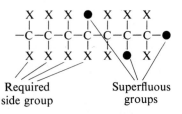

Required side group   Superfluous groups

Naturally if monomers with few superfluous groups are used, cross-linking does not occur so that thermoplastic materials may also be made by the process of condensation.

The degree of cross-linking in a polymer may vary over a very wide range so blurring the boundary between thermosetting and thermoplastic materials. It is convenient, however, to discuss them separately in subsequent sections.

11·3 Mechanical aspects of long chain molecules

Although polymeric molecules have been depicted above as long, straight chains they can flex readily. To see how this happens, consider once again the tetrahedral nature of carbon bonds. Figure 11·6(a) shows the physical relationship between the atoms in the molecule of methane. The molecule of ethane is formed from two CH_3 groups joined together as shown in Fig. 11·6(b) so that the bonds form the same angles with one another as in Fig. 11·6(a).

Although the directions of these bonds are rigidly fixed, it is very easy for either CH_3 group to rotate about the C—C bond. This freedom to rotate about C—C bonds also applies to most polymer chains. In the chain, however, the carbon atoms do not form a straight line (this would violate the directional nature of the bonds) but may be represented by a zig-zag chain:

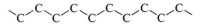

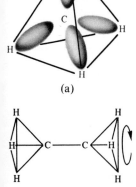

Fig. 11·6 (a) Structure of methane; (b) structure of ethane, showing the possibility of rotation

The hydrogen atoms (or other side-groups) are not shown here for clarity: they are bonded in pairs to each carbon atom in the appropriate tetrahedral fashion. The important feature of the zig-zag is that rotation about a C—C bond *causes the chain to bend*. Thus the zig-zag is only one possible form for the chain; it may bend freely into almost any shape and indeed it will do so continuously at normal temperatures because of the thermal energy present in the material. We may thus picture a polymer as an intertwined net of wriggling chain molecules.

We can now picture what happens when a piece of rubber is stretched, i.e., when a tensile stress is applied between the two ends of a wriggling long-chain molecule. The chain will 'unbend' as the stress increases until it reaches its maximum extension. When the stress is released thermal agitation quickly causes the chain to curl up again. In a piece of solid rubber the tensile stress is transmitted from chain to chain by a small degree of cross-linking between them—this and other aspects of the structure of rubber will be discussed in a later section.

Why, then, do not all polymers show such elastic behaviour? There are two possible reasons why any one material should not do so. Firstly, the structure of the chains may be such as to inhibit free bending. Secondly, the secondary intermolecular bonds may be strong enough to overcome thermal wriggling at normal temperatures (though not at higher ones).

Taking the structural reason first, it should be fairly obvious that if the projecting side groups are sufficiently large they will interfere with one another and inhibit both rotation and the free sliding of chain over chain. Thus polyvinyl chloride is less flexible than polyethylene partly because the chlorine atoms in the former interfere with one another in this way whereas the hydrogen atoms in the latter do not. Polystyrene in which some of the side-groups are benzene rings (see Fig. 11·7) is more rigid then either of these.

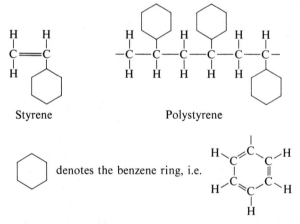

Styrene   Polystyrene

Fig. 11·7 The structure of polystyrene

On the other hand, the flexibility of such structures can be increased by separating the bulky appendages by the inclusion of extra atoms in the backbone of the chain. Suitable units for this purpose are oxygen or sulphur atoms as shown below.

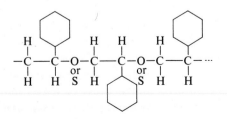

11·4 Mechanical properties of thermoplastic polymers

11·4·1. *Melting point and the glass transition*

At a sufficiently high temperature a thermoplastic polymer is a liquid. In this state it consists of an amorphous mass of wriggling chain molecules. As it is cooled the thermal agitation decreases and at the melting temperature T_m the polymer may crystallize. In the the crystalline state the molecules are all aligned and are packed together in a regular fashion. As may be imagined, this state is not easily achieved for the molecular chains are normally entangled one with another. Thus many polymers will crystallize so slowly that *supercooling*† is readily achieved. These will remain viscous, supercooled liquids until a still lower temperature is reached at which the material *vitrifies*, that is, becomes a glassy solid. The temperature at which this occurs is called the glass transition temperature, T_g.

The mechanical properties of a thermoplastic polymer depend on the values of T_m and T_g relative to room temperature. If both T_m and T_g lie below room temperature the polymer is a liquid. If room temperature lies between T_m and T_g it is either a very viscous supercooled liquid or a crystalline solid. In crystalline polymers the degree of crystallinity is normally limited and the crystallites are separated by amorphous super-cooled regions. Thus polyethylene, which crystallizes readily because of its rather simple molecular structure, is tough and flexible rather than brittle because of the presence of the amorphous fraction. Naturally the ultimate strength, too, depends as much upon the weaker, amorphous part of the solid as on the strength of the crystallites. Polyvinyl chloride on the other hand does not crystallize so readily and is more of a supercooled liquid.

If both T_g and T_m are above room temperature an amorphous polymer is glassy in nature, tending to be brittle. 'Perspex' or 'Lucite' (polymethyl methacrylate) is such a plastic. The temperatures, T_g and T_m, are governed by the strength of the intermolecular forces, just as in inorganic materials, and by the degree of flexibility and length of the chains. Thus polar side-groups such as chlorine and hydroxyl groups favour higher melting and glass transition points because they enhance the strength of the intermolecular bonds. It is clearly possible to design a polymer to have appropriate values of T_m and T_g for a specific application simply by adjusting its chemical construction. However, T_m is generally below 150°C in thermoplastics which have a carbon backbone. Higher melting points are only found in polymers based on a backbone chain of alternating

---

† Supercooling occurs when a material remains liquid below the temperature at which the crystalline solid melts.

silicon and oxygen atoms. As a class these are known as *silicones* or *siloxanes*, and one example, polydimethyl siloxane, is shown in Fig. 11·8.

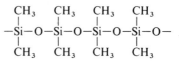

Fig. 11·8 Polydimethyl siloxane

11·4·2. *Viscoelastic behaviour*

In the temperature region between T_m and T_g a linear thermoplastic behaves as a very viscous liquid. Thus the response to a tensile stress is a slow extension of the 'solid', ending in rupture. However, this is not the whole picture since a rapidly applied stress evinces a partly *elastic* response.

This is because the contorted molecules can be instantaneously stretched and only after this do they slowly slide past one another in plastic flow. Thus the strain of a stressed thermoplastic when plotted against time follows the curve shown in Fig. 11·9(a). There is a small initial strain followed by a gradual increase as plastic flow sets in. This combination of viscous and elastic behaviour is termed *viscoelastic*. The time scale over which plastic flow or creep occurs may be minutes, days, or years according to how far the temperature is from T_m and how strong the intermolecular bonds are.

A material which flows without limit is clearly of limited engineering use although electrical insulation and mechanical damping components may be constructed from it provided no tensile strength is required.

Plastic flow may, however, be arrested by joining the molecular chains by strong bonds or by raising the glass transition above room temperature. The latter course produces a rather brittle

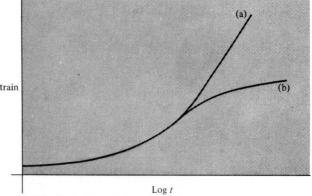

Fig. 11·9 Strain–time curve for a stressed thermoplastic

plastic, of which polystyrene and 'Perspex' ('Lucite') are examples. Linkage of the chains, on the other hand, may be achieved in several ways without losing the toughness which is characteristic of a viscoelastic solid.

11·5   Cross-linking

Cross-links may be formed between molecules in the following ways:

1. By forming a branched network polymer as described in Section 11·1 that is a thermosetting resin. The structure may be likened to that of a fishnet.

2. By introducing short covalent cross-links using divalent atoms (see Fig. 11·10). This also produces a thermosetting resin and may be likened to a fishnet with very long and narrow holes.

3. By the formation of crystallites within the amorphous matrix. These will capture the ends of many molecules, tying them together. We may picture such a solid as a tangled mass of strings, joined together here and there by blobs of glue. Since the crystallites melt at T_m the material retains its thermoplastic nature.

Since thermosetting materials form a separate class they will be left to a later section so that here we concentrate on the properties of partially crystalline thermoplastics. Both types, however, show creep behaviour similar to that shown in Fig. 11·9(b) at sufficiently low temperatures. Since thermoplastics soften rapidly as the temperature rises they are little used for load-bearing articles unless they are highly crystalline.

Polyethylene crystallizes fairly readily and polyvinyl chloride may be induced to do so by introducing a suitable filler, or *plasticizer*, which depresses the glass transition temperature and aids crystallization. It incidentally lowers the rigidity and improves plasticity at moulding temperatures—hence its name. The plasticizers used in polyvinyl chloride are organic liquids such as dioctyl phosphate.

Polyethylene is a particularly versatile material since the degree of crystallinity may be varied widely. The highly crystal-line form called 'high density' polyethylene has much greater rigidity than the 'normal' material, and is much used for liquid containers (buckets, jugs, bowls, bottles) for domestic and industrial use.

The textile fibres polyethylene terephthalate (a copolymer known as 'Terylene' or 'Dacron') and polyhexamethylene adipamide (nylon) are also highly crystalline and display considerable strength as a result. They also exhibit cold-working with consequent improvement in their tensile strength.

11·6 Aging of thermoplastics

Cross-linking by covalent bonds is often accidentally encouraged by the surroundings to the detriment of an article's elastic properties. The aging of polyethylene is a well-known phenomenon and cheap rubber undergoes a similar loss of flexibility with time. The cross-links are formed by oxygen atoms (or by sulphur, if present) under the catalytic action of sunlight. Fig. 11·10 shows how an oxygen atom can join two polymer chains to form a cross-link. This action may be discouraged by including an anti-oxidant, for example a phenol, and an opaque filler such as carbon black to exclude light.

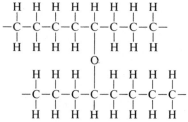

Fig. 11·10 Aging of polyethylene by cross-linking

11·7 Glassy polymers

Polymethyl methacrylate ('Perspex' or 'Lucite') undergoes its glass transition at about 60°C, so that at room temperature it is a glassy solid and does not flow at a measurable rate. Polystyrene is similar: its glass transition occurs at 81°C.

The properties of these plastics are well known and Perspex in particular is much used for producing rigid transparent articles of complex shape. The glassy structure lends it high rigidity coupled with brittleness.

11·8 Mechanical properties of thermosetting plastics

When a component is fabricated from a thermosetting plastic the cross-linkages are most commonly introduced after the object has been hot-moulded from an uncross-linked *resin*, mixed with an appropriate catalytic *hardener*. The chemical reaction which, aided by heat, creates the three-dimensional network is called *curing*. The highly cross-linked polymer which results is rigid and brittle because the cross-links prevent the relative motion of molecular chains. Bakelite (phenol formaldehyde) is such a polymer and was one of the first commercially successful plastics (Fig. 11·11).

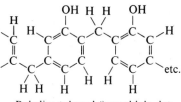

Bakelite (phenol formaldehyde)

Fig. 11·11

Thermosetting plastics do not normally crystallize for the randomly spaced cross-links cannot fit into a regular lattice. In this respect they are somewhat akin to ceramics and phenomena such as slip, dislocations, and their associated effects are unknown in them. The strain–time curve of a thermosetting plastic is similar to that of a crystalline thermoplastic [Fig. 11·9 curve (b)] but is not so dependent on temperature. The covalent cross-links are not broken until relatively high temperatures are reached. One may, for instance, compare the maximum allowable temperature for polyethylene (100°C) with that for Bakelite (200°C). Higher temperatures are permissible for Bakelite if a degree of distortion may be tolerated.

Thermosetting polymers are frequently used with fillers which reduce brittleness or improve their electrical properties. Such composite materials will be discussed later in this chapter.

11·9   Elastomers

Natural rubber is an *elastomer*—a polymer whose limit of elastic extension is very much greater than that of other solids. We have described earlier in the chapter how this property depends upon the fact that contorted molecular chains may be uncoiled by a tensile stress. From the discussion of thermoplastic materials it should be obvious that a small degree of cross-linking is necessary to prevent plastic flow. In natural rubber this is achieved by the process known as *vulcanization*.

The raw material is a viscous fluid called latex which is a linear polymer of isoprene. Figure 11·12 shows two possible structures (two *isomers*) of polyisoprene. The double C—C bond prevents rotation of the monomer units and does not allow the *cis* form to turn into the *trans* form or vice-versa.

Trans-polyisoprene (known as gutta-percha) exhibits a very limited extension under stress because the CH side groups lying on both sides of the chain interfere with one another more than if they were all on the same side.

The liquid form of *cis*-polyisoprene is cross-linked by heating in the presence of sulphur atoms to form a vulcanized rubber with a sulphur content of about 1–2% by weight. If more sulphur is used, a rigid plastic called ebonite is made.

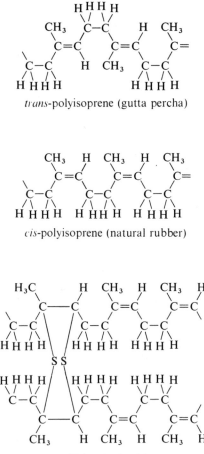

trans-polyisoprene (gutta percha)

cis-polyisoprene (natural rubber)

Vulcanized rubber

Fig. 11·12

As mentioned above, the action of light and oxygen on vulcanized rubber continues the cross-linking process during the life of an article, so a filter such as carbon black is introduced to exclude light from all but the surface, and other chemicals which readily combine with free oxygen atoms may be added.

In spite of such precautions natural rubber eventually deteriorates especially in the presence of oils and greases. As a result, various synthetic rubbers which show greater stability are often used as substitutes. Unfortunately their elasticity and resilience do not quite match those of natural rubber which thus still has its uses. Two examples only of synthetic elastomers are quoted here: polybutadiene and polychloroprene (known as 'Neoprene'). The former, with a similar structure to that of natural rubber and closely similar properties, is commonly used in admixture with the natural product. 'Neoprene', on the other hand, is often used alone but also in blends with other synthetic and natural elastomers. It has exceptional resistance to oxidation and oils.

11·10   Composite materials incorporating polymers

It is often possible to manufacture a superior material by combining two or more substances with complementary properties. For instance in Chapter 10 we explained how the best combination of strength and ductility may be achieved in solids which consist of fibres or precipitated particles embedded in a ductile host material.

Polymers are particularly suitable as components of composite materials as they adhere so readily to the particles, sheets, or fibres of the other component. Some examples of such materials are:

1. 'Tufnol'—a laminated material consisting of layers of woven textiles impregnated with a thermosetting resin. Here the woven material provides great tensile strength while the polymer imparts rigidity.

2. Glass reinforced plastics—G.R.P.—a similar composite which uses glass fibre mat to provide the great strength and the plastic to reduce brittleness.

3. Plasticized polyvinyl chloride—the features of this material have been described above.

4. Plywood—the adhesive used between the cross-grained veneers of natural wood is usually a polymer. The resulting material is less liable to warp and is equally strong in all directions, unlike the straight-grained and cheap softwoods of which it is made.

5. Wood itself is a natural composite material. It consists of fibres of a highly crystalline cellulose polymer (Fig. 11·13)

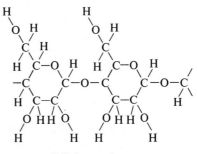

Cellulose polymer

Fig. 11·13

bonded together with amorphous *lignin*, which is composed of carbohydrate compounds. The molecules of cellulose are oriented along the axis of the fibres (that is, along the grain) so that maximum strength is found in that direction. Crystallinity is made possible by the strong bonds arising from the presence of the highly polar hydroxyl side groups.

Note that, unlike the synthetic composites mentioned here, the high-strength component is a polymer, while resilience is provided by the non-polymeric lignin.

6. Vehicle tyres—these use woven cord reinforcement to strengthen the soft energy-absorbing rubber.

7. Expanded polymers—plastics may be made to 'foam' while molten and after cooling they retain large numbers of air cells. The result is a honeycomb structure of great lightness combined with a strength dependent on the amount of plastic per unit volume.

The plastic used may range from a strong thermoset to an elastomer, giving an extremely wide range of available properties. Thus, foamed rubbers allow of even greater elastic deformation than their solid counterparts, while expanded rigid plastics such as polystyrene display great rigidity with lighter weight than can be achieved in solids.

Another useful property is their low thermal conductivity. This is due to the high gas content, gases being very poor conductors of heat compared to solids.

11·11 Electrical properties

Polymers do not display marked magnetic properties for reasons which will become apparent in Chapter 14. Nor do they conduct electricity unless combined with a conducting filler. Indeed, some of the polymers mentioned above are very good insulators. This can be attributed to the fact that all the electrons are strongly bound in the covalent bonds and none is free to conduct a current. In this situation the bulk resistivity is often governed more by the extent to which the plastic will absorb water than by its inherent electrical properties. Thus to select a good insulator we must know which polymers are hydrophobic, that is, do not absorb water.

Polyvinyl chloride, polyethylene, and polytetrafluorethylene are all notable examples of flexible insulators. The last is most useful for its low absorption of power at high frequencies.

Perspex (Lucite) and Tufnol are much used where a rigid insulator is required. Perspex is noted for its resistance to 'tracking', or surface breakdown and subsequent conduction along the track of carbon formed in the breakdown process.

Since some polymers contain polar groups (see Chapter 15) such as chloride atoms, hydroxyl groups, or sulphur atoms they readily become electrically polarized when subjected to an electric field. This results in a high dielectric constant (permittivity) as is explained in Chapter 15. The reader is referred to later chapters for more detailed discussion of the physical principles underlying these properties. It is worth mentioning here that because many plastics contain fillers or plasticizers, their permittivity is artificially increased by the phenomenon known as interfacial polarization. This occurs when a material is composed of two phases having differing electrical properties and it invariably gives rise to big power losses when the material is used to insulate circuits carrying alternating currents.

Problems

11·1 What is the length of a molecule of polyethylene whose molecular weight is 100,000? The C–C bond length is 1·54Å. What length does the chain need to have in order that it may bend to form a full circle?

11·2 Polytetrafluorethylene (ptfe) has the structure shown. Suggest, with reasons, how its properties may differ from those of polyethylene.

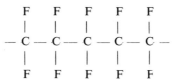

11·3 What is the molecular weight of PTFE when the D.P. is 100,000? How does this compare with the molecular weight of polyethylene with the same D.P.?

11·4 Ebonite is comprised of *cis*-polyisoprene containing about 4% sulphur. Explain why it is a rigid, brittle solid.

11·5 Explain why a glassy thermoplastic is not as hard as an inorganic glass.

11·6 Describe how you would determine experimentally the crystalline melting point of a polymer.

11·7 Suggest why rubber tends to crystallize when it is stretched. (Think of what happens to the shape of the molecules.)

11·8 The monomer vinyl alcohol has the structure shown. Sketch the molecular structure of polyvinyl alcohol and comment on its likely properties.

H   O — H
| |
C = C
| |
H   H

Vinyl Alcohol

11·9 Is it possible to make an elastomer from cellulose (Fig. 11·13)?

11·10 Compare the suitability of wood with that of fibre reinforced thermosetting plastic for furniture manufacture.

Suggest why paint sometimes comes off an article in flakes rather than as a powder.

12 Electrical conduction in metals

12·1 Role of the valence electrons

Many readers will have learned quite a lot about electricity and electric currents without knowing just why it is that some materials will conduct readily while others are insulators which can become statically charged. In the foregoing chapters we have seen how all matter is built of charged constituents—positive and negative—and obviously conduction of electricity must be associated with motion of those charges. Since the protons in the nucleus of an atom are firmly fixed they can only move when the whole atom moves. Now in the case of electrical conduction in metals, we know that no matter is transported when conduction occurs so that the motion of protons cannot be involved and the loosely bound electrons must be responsible for the passage of current. On the other hand, we know in the case of electrolytic conduction that atoms from the electrolyte are released at cathode and anode so that atoms (actually *ions*) are moving during conduction and both positive and negative charges are involved in the process.

For the moment we confine ourselves to metals and we note first that when discussing metallic bonding we pointed out that only the *valence* electrons could be readily removed to take part in bonding. Similarly, we would expect only the valence electrons to be able to take part in conduction. Thus the number of conducting electrons per atom is determined by the atomic structure. Looking at Table 12·1 we can see that copper, silver, and gold have only one such electron per atom, zinc and cadmium have two, while aluminium has three. The univalent metals sodium and potassium are, of course, too reactive to be of engineering importance while if there are more than three valence electrons the character of the bonding changes and with it the electrical properties.

Table 12·1 also gives the resistivities of these elements measured at 20°C and we see that the number of valence electrons alone does not determine its value; it does not follow that more electrons lead to higher conductivity. We must therefore look in more detail at the mechanism of conduction.

12·2 Electrons in a field-free crystal

We begin by considering the behaviour of the valence electrons

Table 12·1

Chemical symbol	Resistivity (Ω-m)	Number of valence electrons per atom
Cu	$1·8 \times 10^{-8}$	1
Ag	$1·6 \times 10^{-8}$	1
Au	$2·4 \times 10^{-8}$	1
Cd	$7·5 \times 10^{-8}$	2
Zn	$6·0 \times 10^{-8}$	2

when there is no potential gradient in the metal and no current is flowing. Having understood this, we then consider how the behaviour is modified when a potential drop is applied across the solid.

The valence electrons in a metal are not only able to leave their parent atoms but they can wander freely through the lattice of ions. One might expect them to collide with each immobile ionic core as was once thought to be the case. But we can easily show that this is not so and that a perfectly regular lattice does not influence the motion of an electron.†

In Chapter 6 we showed that an electron wave of suitable wavelength can de diffracted by a lattice, and in Fig. 12·1(a) we

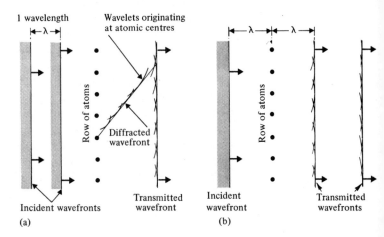

Fig. 12·1 (a) A plane wave is diffracted by a row of atoms when $\lambda < 2d$; (b) when $\lambda > 2d$, diffraction cannot occur

again show a wave diffracted by a lattice. In this diagram the electron wave satisfies the Bragg condition

$$n\lambda = 2d \sin \theta \qquad (12·1)$$

and it is diffracted. In Fig. 12·1(b) the electron wavelength is greater than $2d$ and the condition can no longer be satisfied.

†But see Chapter 13 for exceptions.

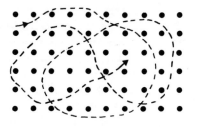

Fig. 12·2 The valence electrons in a metal move along a path which encompasses the whole crystal

As the diagram shows, the only directions in which the wavelets from one row of atoms interfere constructively are forward and backward. When the wavelets from two atomic rows are considered the backward wave disappears through destructive interference and only the forward undeviated wave is left. Thus a valence electron with sufficiently long wavelength (low kinetic energy) is not diffracted by a perfect lattice but moves through it exactly as if the lattice were not there! We can thus calculate the wave-functions for the conduction electrons and determine the expected energy levels by ignoring the presence of the lattice.†

Naturally, in a real metal there are many electrons, all moving in different directions and having different wave-functions, but we begin by asking what are the possible wave-functions for a single electron in the metal. Then we can add the other electrons, taking care not to violate Pauli's principle.

If, as explained in earlier chapters, the electrons are spread uniformly throughout the lattice they cannot be standing waves since in standing waves there are local maxima and minima in the probability distribution. They must therefore be travelling waves since these have uniform intensity everywhere.

Now each electron must be confined within the solid but it is not localized near any atom or group of atoms. It can be regarded as 'orbiting' all the atoms in the solid, moving in a curved path of very large radius so that it is (almost) a plane wave. The path will link every atom in the solid so that the overall path length is very great.

An illustration of this in two dimensions is shown in Fig. 12·2. The path must eventually close upon itself for the same reason as in the hydrogen atom (Chapter 3), namely, that if it were not closed it would eventually lead off to infinity. As in the hydrogen atom, this results in quantization of the energy. Since the path length is very great the spacing between the permitted

† A more exact treatment shows that this statement is not strictly true (see discussion of Brillouin zones in *The Physics of Solids* by Wert and Thomson). However, the elementary approach outlined here gives the main features of metallic conduction, while the more advanced method merely enables the finer details to be understood.

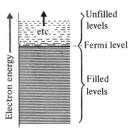

Fig. 12·3 The energy levels available to the valence electrons

energy levels is extremely small; the situation is somewhat akin to that of an electron in the higher levels of a hydrogen atom where the spacing becomes very close because the path is long and contains many wavelengths. However, in this case the energy is wholly kinetic. Because the solid is three-dimensional we find that there are three quantum numbers to describe each wave-function.

A detailed calculation† shows the energy levels to be so closely spaced that the electron can have almost any energy and the range of permitted energy forms an essentially continuous *band*. The energy level diagram is shown in Fig. 12·3, where the spacing between the levels is greatly exaggerated for clarity.

Now, if we add all the other valence electrons one by one to the solid we shall gradually fill these energy levels from the lowest up. Only two electrons, with opposite spins, may have the same set of quantum numbers so that even where several sets of quantum numbers give the same energy there is only a finite number of electrons in each energy level.

Eventually each of the available electrons will have been allocated to an energy level, and all those levels up to a given value, known as the *Fermi level*, will be filled (Fig. 12·3).

We now summarize the important results obtained so far and which will be used in subsequent sections:

(i)   The valence electrons in a metal behave as if the ion cores were absent and the solid were composed of free space.

(ii) The spacing between the energy levels available to the valence electrons is so close that their energy may be regarded for practical purposes as continuously variable.

(iii) In spite of this there is a finite number of available levels, and these are filled by the valence electrons, up to a highest level called the Fermi level. The energy of this level is typically about 4 eV above the lowest level in the band.

12·3   'Electron gas' approximation

Since the kinetic energy of the valence electrons in a metal is almost continuously variable the electrons may be treated as if they were particles in empty space. This important deduction enables us to simplify greatly the calculation of electrical resistivity. Before proceeding with that calculation it is useful to think a little more about the electron in this light. The

---

† The conventional calculation of the energy levels is based on a different treatment of the quantization from that given here. However, the results of the two methods are identical.

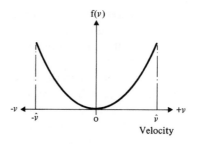

Fig. 12·4 The distribution of velocities among the valence electrons

electrons have a wide range of kinetic energies, as we have seen, owing to Pauli's principle. Thinking of them as particles, we see that this implies a correspondingly wide range of velocities. However, if there is no overall motion of electrons in any direction (that is, no net current) there must be as many electrons moving in one direction as in the opposite direction. The velocity may thus have both positive and negative values but the *average* velocity is zero.

If we plot the fraction, f, of the total number of electrons with a velocity, v, against the value of v, we obtain a curve like that in Fig. 12·4.† All velocities between $-v_{max}$ and $+v_{max}$ are represented. The value $-v_{max}$ is that corresponding to the energy, E_F, of the Fermi level, mentioned in the previous section. By putting $\frac{1}{2}mv^2_{max} = E_F$ it is found that when E_F is 4 eV v_{max} is about $1\cdot2 \times 10^7$ m/s—a very high velocity.

We conclude that the valence electrons in a metal may be pictured as particles moving randomly around at a very high average speed, although also with a wide range of speeds. This suggests that the electrons are very much like a gas of atoms, in which the atoms move about randomly with high thermal energy. The analogy must not be allowed to cloud the fact that the reason why the electrons have high energy (velocity) is quite different from that for the gas atoms. The latter are in rapid motion because of thermal agitation. In the case of the electrons, the effects of thermal agitation are secondary: the velocity of an electron with kinetic energy $\frac{3}{2}kT$ is very tiny indeed compared to the average velocity in Fig. 12·4.

There is one further important effect that must be considered and this is that the electrons fail to 'collide' with the lattice atoms only when the lattice is perfectly regular. If any irregularity is present the moving electron may be scattered and its direction altered. In a real metal there are many such irregularities (see

†The actual shape of this curve is derived from the detailed calculation of energy levels mentioned earlier.

Chapter 7). The most important, however, is due to the vibration of atoms about their mean positions as a result of thermal agitation. Since these vibrations are random they upset the regularity of the lattice structure and the electrons can thus 'collide' with them. We shall see later how it is we know that this is normally the most commonly occurring kind of collision and that the presence of impurity atoms, grain boundaries, and so on, is often of secondary importance.

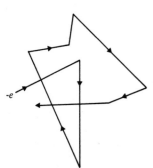

In between these collisions the electrons move undisturbed in a straight line (Fig. 12·5). Their paths are thus random zig-zag patterns threading the lattice.

It should be noted that the speed of the electron may also change when it makes a collision—the loss or gain in its kinetic energy is transferred to or from the vibrating atom with which it collides. However, the overall distribution of electron velocities is still as shown in Fig. 12·4 since for every electron which gains in velocity by collision there is another which loses an equal velocity.

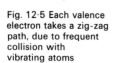

Fig. 12·5 Each valence electron takes a zig-zag path, due to frequent collision with vibrating atoms

We see from the above that the analogy between the electron and a gas is quite close. Just as in a gas, it is possible to define an average path length between collisions, which is usually called the *mean free path*, and a corresponding *mean free time*, namely, the average time spent in free flight between collisions. It is found that the mean free path is about 400 Å in copper at room temperature so that, on average, the electron passes about 150 atoms before colliding with one.

12·4 Electron motion in applied electric fields

Having established the behaviour of electrons in the absence of an electric field, we now turn to the effects produced by applying a field. Since the electrons act between collisions as if they were in free space they are readily accelerated by a potential difference across the crystal, and a large current flows.

The field creates a drifting motion of the whole cloud of electrons in the direction opposite to the field—we often say that the electrons acquire a drift velocity, v_d. This gives rise to an electric current whose magnitude we now calculate in terms of v_d.

Let the current be flowing in the x direction and imagine a cylinder of unit cross section whose axis is also parallel to the x direction (Fig. 12·6). The current flowing down it is equal to the amount of charge crossing any plane, say, plane A, in unit time. This must be just equal to the amount of charge contained in the cylinder in a distance v_d upstream of the plane A. This volume is shown shaded in the figure.

If there are n electrons per unit volume, each of charge, $-e$, the total charge in the shaded volume is just $-nev_d$. They are

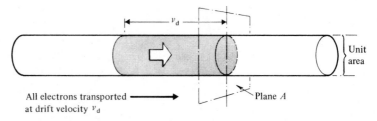

Fig. 12·6 Calculation of the current carried by electrons moving with a drift of velocity v_d

moving in the $+x$ direction but by convention the current is in the $-x$ direction, and therefore has a negative sign. Thus we obtain

$$-J = -nev_d \qquad \text{or} \qquad J = nev_d \qquad (12\cdot2)$$

Since J is the current per unit cross-sectional area it is called the current density. When the current density in a piece of copper is say 1A/cm^2 the drift velocity, v_d is about $7\cdot4 \times 10^{-7}$ m/s (see problem 12·2) so that v_d is only a very small fraction of the average random velocity of the electron gas. It is therefore permissible to assume that the distribution of velocities among the electrons is scarcely affected by the flow of current. All that happens is that each electron aquires an additional velocity equal to v_d; if the original velocity was v the new value is $(v+v_d)$. Note that if v is negative there is a reduction of the magnitude of the velocity. We show in Fig. 12·6 the new graph of $f(v)$ which is the same shape as before but shifted horizontally by an amount v_d in the direction *opposite* to the field $\mathscr{E}$. The size of v_d has been exaggerated to clarify the illustration.

12·5   Calculation of v_d

So far we have assumed that each electron acquires a drift velocity, v_d, without enquiring why it is that the field does not accelerate the electron without limit. To understand this point we must look again at the details of the electron motion. In Fig. 12·6 the path of the electron was depicted as a zig-zag course due to collisions with vibrating atoms. In the interval between collisions the electron experiences a force due to the electric field. Figure 12·7 shows how an electron already travelling in the $-x$ direction is thereby accelerated, while an electron travelling in the $+x$ direction is decelerated. We show in Fig. 12·8 how the path becomes curved when the electron is travelling in some other direction—the curvature is greatly exaggerated for clarity. In each case, however, the force, whose magnitude is $-e\mathscr{E}$, is in the same direction. During free flight the electron is

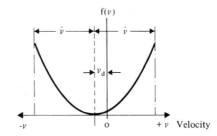

Fig. 12·7 Distribution of velocities among the valence electrons in the presence of an electric field

accelerating; the rate of change of v_d is then equal to the force divided by the electron mass so that in free flight

$$\left(\frac{dv_d}{dt}\right)_{gain} = -\frac{e\mathscr{E}}{m} \tag{12·3a}$$

On the other hand, since the system is in equilibrium we know that, averaged over a significant period of time, $dv_d/dt = 0$. This means that the rate of change of v_d given by Eqn (12·3a) is counterbalanced by the rate at which the drift velocity is lost by collision with the lattice vibrations. The latter can be calculated as follows.

The essentially random nature of a collision is not affected by the fact that the electron has acquired a little extra velocity while in free flight. The direction of the electron velocity immediately after a collision is thus entirely random, as it would have been if the electric field were not present. Immediately after collision, then, the electron's velocity is completely random, and all memory of the drift velocity appears to have been lost.† At the beginning of the next free flight we may assume that $v_d = 0$.

At each collision the change in v_d is thus equal to v_d. If we let τ be the average time spent in free flight between collisions, then each electron collides $1/\tau$ times per second, each time losing a velocity v_d.

The total amount of velocity lost in one second is therefore $v_d \times 1/\tau$, and this is the rate of change of v_d.

Thus the average collision loss is

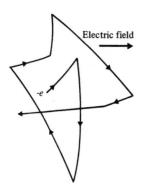

Electric field

Fig. 12·8 The path of each electron becomes curved in an electric field

$$\left(\frac{dv_d}{dt}\right)_{loss} = \frac{v_d}{\tau} \tag{12·3b}$$

Note that, unlike the gain in velocity, the loss occurs only at the time of the collision. The rate of loss is therefore zero for the time, τ, between collisions and very large for the negligible time it

†Note that the *energy* associated with the drift velocity is not lost. It is in part transferred to the lattice and in part gives rise to an increase in the random velocity of the electron.

takes to complete the collision. The average rate is given by Eqn (12·3b).

We have already noted that the total rate of change of v_d is zero so we must add Eqns (12·3a) and (12·3b) and equate to zero, giving

$$\frac{dv_d}{dt} = \left(\frac{dv_d}{dt}\right)_{gain} + \left(\frac{dv_d}{dt}\right)_{loss} = 0$$

Thus

$$\frac{v_d}{\tau} = \frac{\mathscr{E}e}{m} \quad \text{and hence} \quad v_d = \frac{e\mathscr{E}\tau}{m} \tag{12·4}$$

Thus v_d is proportional to $\mathscr{E}$, and it is customary to call the drift velocity in unit electric field the *mobility* of the electron, μ.

Using Eqns (12·2) and (12·4) we obtain the equation for current density:

$$J = \frac{ne^2\mathscr{E}\tau}{m} \tag{12·5}$$

Since the resistivity, ρ, of the metal is just $\mathscr{E}/J$ (see problem 12·3) we have

$$1/\rho = \frac{ne^2\tau}{m} = ne\mu \tag{12·6}$$

where

$$\mu = \frac{e\tau}{m} \tag{12·6a}$$

The two quantities which determine the resistivity are therefore the electron density, n, and the mean free time, τ. Since the electron density cannot vary with temperature the whole of the temperature dependence of ρ must be due to changes in τ. In the next section we shall consider how τ depends on temperature and also how it can be influenced by the defect structure of a metal.

For the moment let us return to consider the implications of the increase in the random velocity which occurs at each collision. Naturally this cannot be a cumulative process and the electrons actually lose some of this extra energy during the collision, transferring it to the vibrating atoms whose vibration therefore increases. This corresponds to an increase in temperature of the solid. Indeed this is just what is observed; the passage of current through a metal causes heating. Since the amount of energy transferred to the lattice increases as v^2_d, the heating effect is proportional to the square of the current.

12·6   Dependence of resistivity on temperature

The reader is probably familiar with the fact that the resistivity of a metal increases linearly with the temperature. It is possible to demonstrate this using the theory outlined in this chapter and we now do so. To simplify the derivation we merely deduce the form of the relationship and the constant of proportionality (that is, the temperature coefficient of resistance) will not be calculated. Although this question appears to be similar to the problem which we considered in Chapter 9, of an atom moving at random in a gas, it is not quite identical and a different approach must be used.

In Eqn (12·4) the only temperature-dependent quantity is the mean free time. We must therefore enquire how this depends upon the amplitude of atomic vibrations and also how the latter depends upon temperature.

The mean free time is connected with the probability of an electron colliding with a vibrating atom: if the probability is small a long time will elapse before a collision occurs and the mean free time will be long. It is in fact a standard result of kinetic theory (see the end of this chapter) that if the mean free time is τ, then the probability of a collision occurring in time δt is $\delta t / \tau$.

It is now necessary to relate the probability of collision to the vibration of an atom. If we assume that the atom oscillates in all directions about its mean position with an amplitude, a, then it sweeps out a spherical volume of radius a during its motion. Now the line of flight of the electron may or may not pass through this volume. The possibility of this happening is just proportional to the area which is presented to the oncoming electron (Fig. 12·9) and this will be equal to the probability of a collision occurring. The collision probability is thus proportional to a^2 and the mean free time to $1/a^2$, that is,

$$\tau \propto a^{-2} \tag{12·7}$$

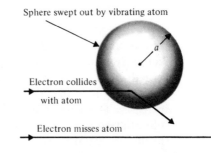

Fig. 12·9 Collision of an electron with a vibrating atom

We must now find how the amplitude a depends upon temperature. Since the atom is in thermal equilibrium with its surroundings we know from kinetic theory that its thermal energy is $\frac{3}{2}kT$, where k is Boltzmann's constant and T is the absolute temperature. We can put the energy in terms of the amplitude of oscillation as follows. The atom is oscillating back and forth against the 'spring' of the bonds which join it to its neighbours. If its instantaneous deviation along the x direction from its rest position is x, then x varies sinusoidally with time:

$$x = a \sin 2\pi ft \qquad (12\cdot8)$$

where f is the frequency of oscillation.

The kinetic energy of the atom, whose mass is M, is just

$$\tfrac{1}{2}M\left(\frac{\mathrm{d}x}{\mathrm{d}t}\right)^2 = \tfrac{1}{2}M(a^2\,4\pi^2 f^2 \cos^2 2\pi ft) \qquad (12\cdot9)$$

The total energy is partly kinetic energy and partly energy stored in the 'spring'. Each of these varies with time, although by the law of energy conservation their sum remains constant. When the displacement, x, is zero there is no elastic energy in the spring so the energy is wholly kinetic. The total energy of oscillation, E, is given by Eqn (12·8) when $x = 0$, that is, when t = 0. Thus

$$E = 2\pi^2 M f^2 a^2$$

As stated earlier, the energy is proportional to temperature so that we have the relationship we are seeking:

$$a^2 \propto E \propto T \qquad (12\cdot10)$$

Using both Eqn (12·10) and Eqn (12·7) we find that the mean free time is inversely proportional to temperature, thus

$$\tau \propto 1/T \qquad (12\cdot11)$$

Finally, looking back to Eqn (12·4), which relates the resistivity to τ, and using Eqn (12·11) we see that

$$\rho \propto \frac{1}{\tau} \propto T \qquad (12\cdot12)$$

This is the result we set out to prove.

12·7   Structural dependence of resistivity

The relationship (12·12) implies that the resistance of a piece of metal tends towards zero at a temperature of absolute zero. When measurements are made at low temperatures a deviation from this law is observed: the resistivity tends to a finite value

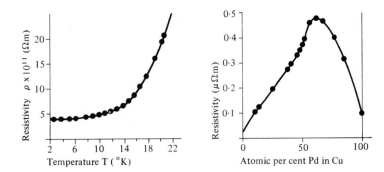

Fig. 12·10 Resistivity of sodium at low temperature

Fig. 12·11 Resistance of copper–palladium alloys

(Fig. 12·10) as the temperature is lowered. This is because the atomic vibration becomes so small at these temperatures that collisions with vibrating atoms become less important than those with crystal defects such as impurities, grain boundaries, and so on. Since these are fixed in number the mean free time no longer depends so strongly on temperature.

This result suggests that it might be possible to increase the resistance of a metal at ordinary temperatures and at the same time reduce its temperature coefficient of resistivity, by introducing many defects. This is achieved in some disordered solid solutions whose resistance may be much higher than that of either constituent. An example of this is shown in Fig. 12·11. The copper–nickel alloys (for example, 'Constantan') which are used for making electrical resistances show similar behaviour, though not entirely for the same reason.

On the other hand, the behaviour is quite different in a two-phase alloy in which the structure at intermediate compositions is eutectic. In this case the material is essentially a mixture of two phases with different resistivities, and from an electrical viewpoint it may be regarded as a complicated network of resistances in series–parallel with one another. It can be shown that the resistance of such a structure varies linearly with the volume fraction of either constituent. Thus, if the resistivities of the two phases, α and β, are ρ_α and ρ_β then the resistivity of the two-phase material is

$$\rho = \rho_\alpha V_\alpha + \rho_\beta V_\beta$$

where V_α and V_β are the volume fractions of the α and β phases.

*12·8   Collision probability and mean free time of electrons

In this section we prove the relationship which was used in deriving Eqn (12·7).

The probability of an electron undergoing collision within time δt is proportional to δt, so we may let it equal $p\delta t$. Take a group of N_0 atoms at time $t = 0$. Some of these will collide only a short time later, others will be in free flight for longer. To calculate the *mean* free time we must know how many are left (i.e., have not collided) after an arbitrary time t; for the moment we assume that there are N of these.

Between times t and $(t+\delta t)$ a further small number, δN, will have collided. This number δN will be equal to the number N left in the group, multiplied by the probability of collision, $p\delta t$, for each electron, that is

$$\delta N = -Np\,\delta t \tag{12·13}$$

The minus sign enters because δN represents a decrease in N.

We now find the relationship between N and t by integrating Eqn (12·13) between times $t = 0$ (when $N = N_0$) and t (when $N = N$), thus,

$$\int_{N_0}^{N} \frac{dN}{N} = -\int_0^t p\,dt$$

Performing the integration we find that

$$N = N_0 \exp(-pt) \tag{12·14}$$

Now Eqn (12·13) tells us the number of electrons, δN, which collide between t and $(t+\delta t)$, and which therefore have a time of free flight equal to t. The average flight time, τ, is given by

$$\tau = \frac{1}{N_0} \int_0^{t=\infty} t\,dN$$

with the help of Eqn (12·13) this becomes

$$\tau = -\int_0^\infty \frac{N}{N_0} pt\,dt.$$

Using Eqn (12·14) this may be integrated by parts as follows

$$\tau = -t \exp(-pt)\,dt$$

$$= -\frac{1}{p}\left[t\exp(-pt)\right]_0^\infty - \frac{1}{p}\int_0^\infty p\exp(-pt)\,dt$$

$$= -\frac{1}{p}\left[t\exp(-pt)\right]_0^\infty + \left[\frac{1}{p}\exp(-pt)\right]_0^\infty$$

Since the first term is zero at both $t = 0$ and $t = \infty$ we finally discover that

$$\tau = \frac{1}{p}$$

so that the collision probability is just $1/\tau$ per unit time.

Problems

12·1 Show that the following two definitions of the resistivity ρ are equivalent:

(a) Resistance of a bar $= \rho \times \dfrac{\text{length of bar}}{\text{cross-sectional area of bar}}$

(b) $\rho = \dfrac{\text{Electric field}}{\text{Current density}}$

12·2 The density of copper is $8·93 \times 10^3$ kg/m³. Calculate the number of free electrons per cubic metre and hence deduce their drift velocity when a current is flowing whose density is 1 A/cm².

12·3 The Fermi energies and interatomic distances in several metals are given below. Check in each case that the electrons of highest energy have a wavelength greater than twice the distance between atomic planes, and hence are not diffracted by the lattice.

Metal	Na	K	Cu	Ag
Fermi energy (eV)	3·12	2·14	7·04	5·51
Interplanar spacing (Å)	3·02	3·88	2·09	2·35

12·4 What is the kinetic energy of an electron which has a wavelength just short enough to be diffracted in each of the metals listed in Problem 12·3?

12·5 From the data given in Table 12·1 and Appendix 3, calculate the mobility of an electron in each metal at room temperature.

12·6 What is the mean free time between collisions in the metals Cu, Ag, and Au? Their resistivities are given in Table 12·1 and their densities are respectively 8·93 gm/cm³, 10·5 gm/cm³, and 19·35 gm/cm³.

12·7 Deduce the maximum random velocity of electrons in Cu, Ag, and Au if their Fermi energies are respectively 7·04 eV, 5·51 eV, and 5·51 eV. Using these velocities and the mean free times obtained in Problem 12·6, find approximate values for the electron mean free path in the three metals.

12·8 When about 1 atomic percent of a monovalent impurity is added to copper, the mean free time between collisions with the impurities is about 5×10^{-14} s. Calculate the resistivity of the impure metal at room temperature.

Using the electron velocity deduced in Problem 12·7, find the mean distance between collisions with impurities. Compare this with the average distance between the impurity atoms and explain why the two distances are different.

12·9 If the electron mean free path were independent of temperature (as a result, say, of a large impurity content) and the electrons behaved as if they constituted a perfect classical gas, show that the resistivity would be proportional to $T^{1/2}$.

12·10 As explained in Chapter 11, polyvinyl chloride normally contains several additives which improve its mechanical properties. What effect would you expect each to have on the electrical resistivity of p.v.c.?

12·11 In Chapter 8 the changes in a metal which result from plastic deformation were described. Suggest what changes you would expect in the conductivity of copper as a result of cold working. How could you restore the resistivity to its original value?

13 Semiconductors

13·1 Introduction

If one examines the electrical resistivities of the elements, particularly of the elements in Groups I B, II B, III B, and IV B, it will be seen that they tend to increase with increasing valency. At first sight, this is surprising since increasing valency means an increasing number of loosely bound electrons in the outer shells of the atoms, and an increase in the number of electrons available might be expected to lead to an increase in electrical conductivity, especially in view of Eqn (12·6). The resistivity of copper (Group I B) is about $1·7 \times 10^{-8}$ ohm-metre at room temperature while that of germanium (Group IV B in the same period of the Table) is in the range 10^{-3} to 10^4 ohm-metre, depending on its purity. This phenomenon is directly related to the interatomic bonding in these materials.

13·2 Bonding and conductivity

It will be recalled that the bonding in a metal is visualized in terms of a fixed array of positive ions held in position by a 'sea' of negative charge. The sea consists of the valence electrons which are distributed throughout the solid, i.e., they are shared by all the atoms. In the Group IV materials covalent bonds are formed whose distinguishing feature is that the electrons are *localized* in the bonds. This means that the wave-function of a bonding electron is confined to a region of space between adjacent atoms, as depicted in Figs. 5·9 and 5·12. In fact, the maximum probability density for the bonding electrons is just midway between the atoms. We show this in the diamond lattice in Fig. 13·1, where for clarity the three-dimensional structure has been 'unfolded' and drawn in two dimensions. Each bond contains two electrons with opposite spin directions—more than this is not possible owing to Pauli's exclusion principle.

On must not think, however, that these electrons are immobile, condemned forever to stay in the same bond. The probability distributions are smeared out to such an extent that those of neighbouring bonds overlap slightly, and this makes it possible for a pair of electrons in adjacent bonds to change places with one another. Indeed, all the electrons are continually doing so, moving through the network of bonds at great velocity.

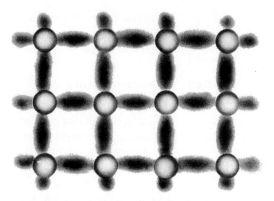

Fig. 13·1 A two-dimensional representation of the covalent bonds in a crystal of C, Si, or Ge; the electron clouds are concentrated between the atoms

In spite of this rapid motion, Pauli's principle still ensures that at any one time there are no more than two electrons in any particular bond.

Now although the electrons are moving through the network of bonds, they are not strictly to be regarded as travelling waves. Travelling waves have a constant intensity at every point, while we know that in covalent bonds the electron clouds are concentrated midway between the atoms. The electron density (Fig. 13·1) is more like the distribution in a set of standing waves, with nodes at the ion centres and antinodes between them as suggested in Fig. 13·2.

The electron waves are therefore standing waves, not running waves. However, this does not mean that the electrons cannot change places with one another, since each standing wave is composed of two running waves travelling in opposite directions. Because the amplitudes of the two oppositely travelling waves are equal, there is exactly as much charge being transported in one direction as the opposite one. In this way, the charge density at each point remains unchanged, and Pauli's principle is

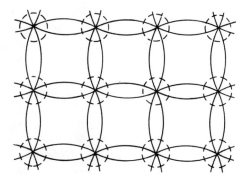

Fig. 13·2 The electron distribution in the diamond structure is like a system of standing waves

satisfied. From this we deduce that no net flow of charge in any direction is possible, i.e., a covalent crystal cannot carry an electric current.

13·3   Semiconductors

The above argument suggests that a crystal with covalent bonding should be a good insulator, and in some cases this is substantially correct. Our prototype covalent solid, diamond, has a resistivity of 10^{12} Ωm at room temperature, and polymers such as polyethylene and polyvinyl chloride have resistivities of about 10^{15} Ωm in their pure state. However, two elements, silicon and germanium, both of which are of great technological importance, have much lower resistivities. Why is this? If we compare the bond strengths of diamond, silicon, and germanium, we find that they decrease as the atomic weight increases. (This has already been discussed in Section 6·6.) It is thus easier to break the bonds in germanium than in silicon, while it is hardest of all to break them in diamond. If the bond is sufficiently weak, the thermal vibration of the atoms may occasionally be sufficient to sever a bond, even when the temperature is not very high. Naturally, the higher the temperature, the more likely it is that any one bond will be broken, and hence the more bonds will be broken at a given time.

Now breaking a bond does not affect the atoms which it joins together—each still has three bonds to keep it in place. But breaking a bond does entail releasing one of the electrons which form it, as shown in Fig. 13·3. Thus at any temperature a small proportion of the large number of bonding electrons will be set free. Being no longer confined in the bonds, these electrons may wander freely through the lattice and conduct an electric current. At room temperature the freed electrons in a piece of silicon

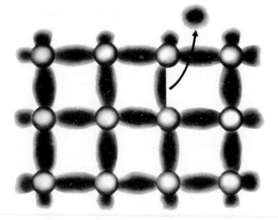

Fig. 13·3 Thermal vibration causes an electron to be freed from one of the bonds

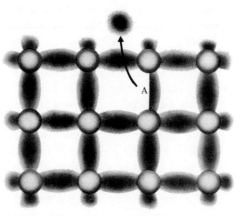

Hole is created at A

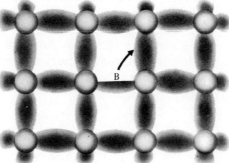

Electron moves from neighbouring bond to fill hole, which now appears at B

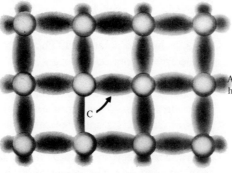

Another electron moves to fill hole, which is now at C

Fig. 13·4 As the bonding electrons move about, the hole is filled and appears to move

constitute only about one in 10^{13} of the total number of valence electrons. This is in complete contrast to the situation in a metal, in which all the valence electrons are free to conduct current.

The number of free electrons is essentially fixed at a given temperature, although the process of bond rupture by thermal agitation is continuous. The rate at which electrons are freed is exactly balanced by the rate at which those already free are trapped by an empty bond and lose their freedom.

We shall postpone the calculation of the conductivity generated by the presence of free electrons to the next section, for we must now take note of a curious phenomenon which occurs in the bonds.

The freed electrons leave behind them gaps in the bonds, called holes. Since the bonded electrons are in constant motion, these holes are soon filled by nearby electrons. This necessarily causes the holes to appear in the bonds vacated by the itinerant electrons, and the holes appear to move (Fig. 13·4). In fact the holes move about in much the same random way and at the same velocity as the electrons in the filled bonds.

We shall now show that the presence of these holes enables the bonded electrons to carry a current in such a way that it appears as if it is conducted by positive charges carried by the holes!

13·4   Conduction by holes

We have seen that the hole moves by virtue of an electron jumping into it from an adjacent bond. We have also stressed that, when all the bonds are full, the electron distribution is not affected by an electric field. However, when a field is applied and a hole is present, the probability of an electron jumping into the hole from an adjacent bond will be increased if its jump is assisted by the field; conversely, the probability of a jump against the field is lowered. Thus, if the field favours the motion of electrons from left to right (say) in Fig. 13·4, the hole will tend to drift from right to left and thus its motion is influenced by the field. Furthermore, its drift direction in the field is opposite to that of electrons and so it behaves as if it carried a *positive* charge. Its speed of movement between bonds may be divided into two components: a random velocity arising from the random electron motion; and a drift velocity which comes from the fact that the field makes the electron velocity slightly greater in one direction than in the other. Thus, a hole experiences an acceleration in an electric field equal but opposite in direction to that of a free electron, that is, it behaves like a free electron with *positive* charge. Its drift velocity is limited by

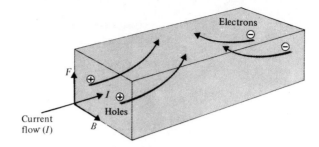

Fig. 13·5 The forces on electrons and holes moving through a magnetic field

collisions in just the same way as is the drift velocity of an electron, because its movement is determined by the drift of real electrons in the opposite direction.

Since the drift velocity is a direct result of the application of an electric field, the hole makes a direct contribution to the conductivity of the crystal.   If there are p holes present the current they carry is p times the current carried by one hole: it is as if there were $\dot{p}$ positively charged 'electrons'.

The validity of this point of view has been demonstrated in experiments in which both magnetic and electric fields are applied. Consider a charged particle which is accelerated to a drift velocity v in an electric field $\mathscr{E}$. If a magnetic flux B is also present, lying perpendicular to the direction of $\mathscr{E}$, then the magnetic force F on the particle, as given in Chapter 1, is $F = Bev$ in a direction perpendicular to both v and B (Fig. 13·5). Note that if the sign of e is changed, the direction of motion due to the electric field is reversed, i.e., v becomes negative. The force F, however, remains in the same direction because *both* e and v have changed sign. The particle, whether hole or electron, tends to move in the direction of F and by detecting this transverse motion electrically it is possible to determine the sign of its charge [Fig. 13·5(b)]. Such experiments when carried out on semiconductors confirm the existence of positive holes.

13·5   Energy bands in a semiconductor

We have seen that there are two kinds of electrons in a semi-conductor: those in the bonding system and those that are 'free'. The free electrons move about in the crystal and contribute to electrical conductivity in just the same way as the free electrons in a metal. They can be treated as plane waves and their energies must be quantized in a large number of closely spaced energy levels as were the electrons in a metal. The band of energies

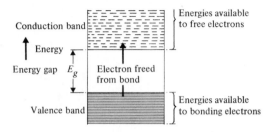

Fig. 13·6 The energy diagram for a semiconductor, showing the energy gap

formed by these levels is called the *conduction band*. Since the electrons in the covalent bonds also move about at varying speeds, they also have a range of permitted energies. Detailed consideration of the quantization conditions shows that these levels are also closely spaced and form an energy band called the *valence band*.

The energies of the levels in the valence band must all be below those of the conduction band, for we have seen that an electron has to acquire energy from the thermal vibrations in the lattice in order to leave a bond and become free. This situation is shown on the energy band diagram in Fig. 13·6. There is, in fact, a minimum value for the amount of additional energy which a bonding electron must acquire in order to leave the bond. The absolute minimum for this energy is that which the most energetic bonding electron needs in order just to become free, with no kinetic energy left over to carry it away. This will be represented in the energy level diagram by a gap between the highest level in the band of energies occupied by the bonding electrons, that is, the valence band, and the lowest level in the conduction band, as shown in Fig. 13·6. This range of energies which is prohibited to the electrons is called the forbidden energy gap, or often just the energy gap, and is a direct measure of the amount of energy an electron needs to leave the bonds. It is different for different materials. In germanium it is 0·65 eV, in silicon 1·15 eV, while in diamond it is 5·3 eV.

The arrow in Fig. 13·6 represents the transition of an electron from the valence band to a state in the conduction band, i.e., from a bond to freedom. This leaves vacant a state in the valence band, i.e., a 'hole'. A downward arrow would represent an electron falling back into a lower level, and hence recombining with a hole.

This diagram gives us another way of looking at conduction by electrons. Clearly, electrons in the conduction band can accelerate readily because empty, higher energy levels are available to accommodate them. If a hole (i.e., a vacant state) exists at the top of the valence band, an electron in a lower level can be given energy to fill it and the accelerating electron carries a current. On the other hand, if there were no vacant levels in the valence band no electrons could be accelerated.

13·6 Excitation of electrons

When an electron near the top of the valence band (i.e., in a bond) collides with a vibrating atom, it can absorb some of the atom's energy and jump into the conduction band. This is just another (more accurate) way of describing how thermal vibrations can break a bond and create an electron–hole pair.

According to the thermodynamic arguments presented in Chapter 9, we might expect that the probability of an electron acquiring a thermal energy E to be $\exp(-E/kT)$. In the present case, however, this expression is not exactly correct, because the electron must satisfy Pauli's principle, which in effect says that the electron cannot acquire the energy E unless there is a vacant level at the appropriate energy. For instance, in the present case nearly all the valence levels are full, so that jumps of electrons from the lowest levels are very restricted.

When Pauli's exclusion principle is taken into account we find that the probability of an electron acquiring enough energy to cross the energy gap E_g is approximately proportional to

$$\exp\left(\frac{-E_g}{2kT}\right)$$

The actual number n of electrons per unit volume of the crystal which are in the conduction band is obtained by multiplying this probability by the number of energy levels per unit volume in the conduction band which may accept an electron. This we shall just call N_c.   Thus

$$n = N_c \exp\left(-E_g/2kT\right) \tag{13·1}$$

Since the number of holes, p, is equal to n we also have

$$p = N_c \exp(-E_g/2kT) \tag{13·2}$$

Now N_c is a constant at a given temperature—it is approximately equal to $2 \cdot 5 \times 10^{25}$ per cubic metre at $300°K$. It is nearly independent of the semiconductor and is so weakly dependent on temperature that its variation is completely overshadowed by the exponential in Eqns (13·1) and (13·2) and its temperature dependence may be ignored.

The electrons in the conduction band and the holes in the valence band may carry current in a similar way to that described for metals in Chapter 12. Thus we can use Eqn (12·2) to write down the current density J_n carried by n free electrons per cubic metre whose drift velocity is v_n:

$$J_n = nev_n$$

Note that J_n has a positive sign, since both v_n and e are negative. Similarly the current density J_p due to the motion of holes in the valence band may be written in terms of their drift velocity $+v_p$

$$J_p = pev_p$$

where p is the number density of positive holes. The total current density J is then the sum of these

$$J = nev_n + pev_p \tag{13·3}$$

The number of free electrons n and the number of holes p per unit volume are equal† and given by Eqn (12·1) so that we may write the current density

$$J = N_c (ev_n + ev_p) \exp\left(-\frac{E_g}{2KT}\right) \tag{13·4}$$

Assuming that both the free electrons and the holes behave like the free electrons in a metal, we may use the results of Chapter 12. Thus v_n and v_p will both be proportional to the electric field $\mathscr{E}$ (Eqn 12·4) and we may divide Eqn (13·3) by $\mathscr{E}$ to obtain the conductivity:

$$\frac{J}{\mathscr{E}} = \sigma = N_c (e\mu_n + e\mu_p) \exp\left(-\frac{E_g}{2kT}\right) \tag{13·5}$$

Where μ_n and μ_p are the drift velocities of electrons and of holes respectively in a unit electric field. As before we call μ_n and μ_p the *mobilities* of electrons and of holes respectively. In general they will not be equal to one another, but we may expect them both to vary inversely with temperature as we showed for electrons in Chapter 12. However, the exponential in Eqn (13·5) varies more rapidly with temperature and completely dominates the temperature dependence of the whole expression. Neglecting the temperature dependence of μ_p and μ_n, we see that σ rises approximately exponentially with temperature, in marked contrast to metals, for which it falls with rising temperature. The reason is that for semiconductors the *number* of available charge carriers grows exponentially as the tempera-

---

† They are not always so; see Section 13·7.

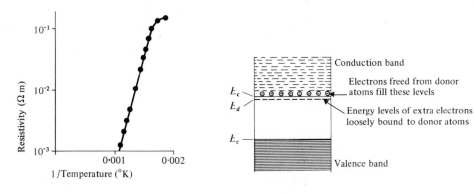

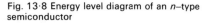

Fig. 13·7 The dependence of the resistivity of silicon on temperature

Fig. 13·8 Energy level diagram of an *n*–type semiconductor

ture rises, while for metals it is roughly constant. For semi-conductors, the rising temperature gives more and more valence electrons enough energy to jump into the conduction band, freeing them and the corresponding holes to conduct an electric current.

The experimental values of the conductivity of silicon measured at different temperatures T are plotted on a log scale against $1/T$ in Fig. 13·7. Exactly as predicted by Eqn (13·5), the points lie very closely on a straight line, the slope of which can be seen by differentiation to be equal to $E_g/2k$. Using the value for Boltzmann's constant k we can find from the graph that for silicon $E_g = 1·15$ eV, as stated earlier.

There are more accurate ways than this of determining energy gaps, and the values for several materials are given in Table 13·1. Among these, silicon and germanium have been the materials most used for transistors, diodes, etc., though now-adays some compound semiconductors are being used for special purposes. Gallium arsenide (GaAs) with an energy gap of 1·45 eV is used in an infra-red light source, while indium antimonide (InSb, $E_g = 0·18$ eV) is used for detection of infra-red radiation and in magnetic field detectors. Cadmium sulphide (CdS, $E_g = 2·45$ eV) is a detector of visible light.

Among these compounds a wide range of properties is available—different energy gaps, different electron and hole mobilities—which makes each suitable for a different application. In this chapter we shall just consider diodes and transistors, whose operation depends upon the controlled introduction of very small amounts of impurities into an initially very pure crystal.

Table 13·1   Energy gaps, electron densities, and conductivities

Material	E_g(eV)	Calculated value of n at 300°K (m^{-3})	Measured value of σ at 300°K (Ω^{-1} m^{-1})
Copper	None	8.5×10^{28}	5×10^7
Aluminium	None	1.8×10^{29}	3.3×10^7
Gray tin	0·08	5.4×10^{24}	10^6
Germanium	0·65	9.3×10^{19}	2.2
Silicon	1·15	6×10^{15}	5×10^{-4}
Diamond	5·3	1.1×10^{-20}	10^{-12}†
LiF	11	2×10^{-70}	10^{-11}†

†Conductivity in these materials does not occur by the motion of electrons since other processes occur more readily.

13·7   Impurity semiconductors

Each germanium or silicon atom in the semiconducting crystal has four valence electrons which it shares with its neighbours to form four covalent bonds. From another viewpoint, these electrons fill energy levels in an energy band (the valence band) so that the band is just filled at the zero of absolute temperature.

Suppose that a pentavalent atom (Group V) were to be substituted for one of the silicon atoms. An arsenic or antimony atom would be suitable, but a phosphorus atom would fit better then either of these into the silicon lattice, since its atomic radius is closely similar to that of silicon. Four of the five valence electrons belonging to this atom are required to form the four covalent bonds with the neighbouring silicon atoms. The fifth electron is bound to the atom by the extra positive charge on the nucleus, but, as we shall shortly show, it requires very little energy to break this bond, and at room temperature the thermal energy in the lattice is more than sufficient to do so. The extra electron is then free to move through the silicon lattice.

Let us now build up the same picture using the energy band model. The freed electron has been excited to an empty state in the conduction band. Since the energy required to remove it from its parent nucleus was very small, it must have come from an energy level *just below* the bottom of the conduction band—in the energy gap! (Fig. 13·8.) Thus the introduction of the Group V atom has created a new energy level, near to the conduction band, which is normally vacant at room temperature because the electron is donated to the conduction band—the impurity is called a *donor* for this reason.

In order to indicate that the donors are localized, we make the horizontal axis of the band diagrams represent distance through the crystal: the donor levels are then shown as short lines.

We may show that the binding energy of an electron in this level is small by regarding the extra electron and the excess

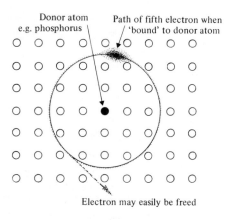

Donor atom e.g. phosphorus

Path of fifth electron when 'bound' to donor atom

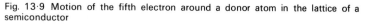

Electron may easily be freed

Fig. 13·9 Motion of the fifth electron around a donor atom in the lattice of a semiconductor

positive nuclear charge as if they were like a hydrogen atom. In both situations a single, negatively charged electron moves in the electric field of a single, fixed positive charge (Fig. 13·9). But in the present case the electron moves through a silicon lattice which we may regard approximately as a medium with a large relative permittivity (dielectric constant) ε_r. The expression derived in Chapter 3 for the hydrogen atom can be used now to derive the binding energy, but we must replace ε_0 by $\varepsilon_r\varepsilon_0$ giving the result

$$\text{Binding energy} = \frac{me^4}{8\varepsilon_0^2\varepsilon_r^2 h^2} \tag{13·6}$$

This is the energy difference between the donor level and the bottom of the conduction band, i.e., it is equal to $(E_c - E_d)$ (Fig. 13·8).

Since the relative permittivity of silicon is 11·7, the binding energy is smaller than the ionization energy of hydrogen by a factor $11\cdot7^2$, so that

$$(E_c - E_d) = \frac{13\cdot6}{11\cdot7^2} = 0\cdot1\,\text{eV}$$

Naturally this result is only approximate [$(E_c - E_d)$ is actually nearer 0·05 eV] but it serves to show that the binding energy is indeed small compared with the width of the energy gap, which in silicon is 1·15 eV. The corresponding values for germanium are $(E_c - E_d) = 0\cdot01$ eV and $E_g = 0\cdot65$ eV.

The density of free electrons in *pure* silicon, n_i, (often called the *intrinsic* value) can be calculated from Eqn (13·2). Using the value for N_c given earlier, the density of electron–hole pairs at $T = 300°$K is found to be

$$n_i = N_c \exp(-E_g/2kT)$$
$$= 2 \cdot 5 \times 10^{25} \times \exp(-1 \cdot 15/0 \cdot 026 \times 2)$$

Thus

$$n_i = 6 \cdot 3 \times 10^{15} \, \text{m}^{-3} \tag{13.7}$$

On the other hand there are 2×10^{29} valence electrons per cubic metre and about 5×10^{28} atoms per cubic metre in solid silicon. This confirms our earlier statement that only a few electrons are thermally excited across the band gap.

If now only one silicon atom in 10^{10} were replaced by a donor impurity, the density of donors would be about $5 \times 10^{18}/\text{m}^3$, and the same number of extra electrons would be available to conduct an electric current. This is far in excess of the number of carriers in the intrinsic material, so the conduction process is dominated by the presence of the electrons which are called the majority carriers. Since the current carriers are almost exclusively negative, the material is termed an n-type impurity semiconductor.

We have glossed over one point in the argument. Not all the donor electrons appear in the conduction band. Just as in the case of intrinsic electrons, the number of electrons in the conduction band must be calculated from the probability of a donor electron being excited across the gap between the donor level E_d and the conduction band edge E_c. This probability is calculated in more advanced textbooks (see Bibliography) with the result that, at room temperature, nearly all the donor atoms are found to be ionized, unless the density of donors becomes too large (greater than about $10^{25}/\text{m}^3$ in germanium).

For present purposes we may therefore treat a piece of n-type semiconductor containing N_d donors per unit volume as a conductor in which the free electron density is approximately equal to N_d.

13·8 Impurity semiconductors containing Group III elements

By contrast with the action of Group V elements, small traces of a Group III element such as boron or aluminium create a deficiency of electrons. Each impurity atom has only three valence electrons, while four are required for bonding. The incomplete covalent bond contains a hole which may, if it escapes from the impurity atom, take part in conduction.

The escape of the hole occurs by the capture of a valence electron from a neighbouring silicon atom, i.e., from the valence band on the energy diagram. For this reason the Group III impurity is called an acceptor atom. The captured electron now has a slightly higher energy than the other valence electrons,

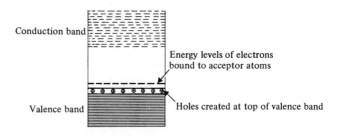

Fig. 13·10 Energy diagram of a *p*-type semiconductor

since it has upset the neutrality of the impurity atom, which has now become a negative ion. We therefore allocate to the captured electron an energy level called an acceptor level just above the top of the valence band (Fig. 13·10). Since the energy gap between the top of the valence band and the acceptor level is very small, electrons from the valence band are readily excited across it and most of the acceptor levels will be filled at normal temperatures. A corresponding number of holes is created in the valence band and these may conduct an electric current.

As with electrons in *n*-type silicon, the number of holes created by only a small concentration of acceptors may greatly exceed the number of holes in intrinsic silicon [given in Eqn (13·7)] and conduction thus occurs primarily by the motion of these positive holes. The material is called a *p*-type impurity semiconductor, and in contrast to *n*-type material the majority of carriers are now holes.

The energy gap between the valence band and the acceptor levels may be calculated in a similar way to that for donor levels. Since the captured electron gives an acceptor atom a charge $-e$, it has an 'attraction' for holes. So a hole can become bound to an acceptor as if it were an 'inverted' hydrogen atom—a positive charge orbiting a fixed negative charge. Equation (13·6) can again be used to obtain an approximate value for the binding energy, which is thus about 0·05 eV in silicon.

Many other impurities may be put into silicon or germanium to create new energy levels at almost any point in the energy gap, but the Group III and Group V elements are by far the most important for practical purposes. Many kinds of lattice defects can also act as donor or acceptor levels, or as traps for current carriers. It is therefore important to prepare the materials for diodes and transistors with the greatest care, to reduce both unwanted impurities and lattice defects below very low levels. As a result of the development of new technologies for doing this, silicon and germanium have become two of the purest obtainable elements.

13·9   The *p–n* junction: a rectifier

A piece of semiconductor in which a section of *p*-type material is joined to one of *n*-type material is called a *p–n* junction. It is on the properties of this junction that the semiconductor diode and the transistor depend. Such a junction is not usually made by joining two separate pieces of material, but by taking a single crystal of silicon and introducing donor and acceptor impurities, perhaps by diffusion, into the appropriate parts of the solid. We shall now consider the electrical properties of such a junction.

Let us imagine what happens to the electrons and holes in two pieces of semiconductor, one *p*-type and one *n*-type, as they are placed together. The initial distribution of holes and electrons is shown in Fig. 13·11(a).

On one side of the junction is a material with many electrons freely wandering through the lattice, while on the other side there is a material with many free holes. It is natural that some of these excess carriers will wander across the junction into the 'foreign' material, just as two gases will diffuse into one another. But, unlike gases, these electrons and holes can combine self-destructively with one another. That is, the excess electrons wandering into the *p*-type crystal can fill some of the holes (which are just vacant energy levels) so that neither can carry current. Similarly, holes diffusing into the *n*-type region recombine there with the electrons. Obviously, this only occurs near to the junction since the carriers cannot move far across the junction without recombining. The resulting distribution of carrier densities is shown in Figs. 13·11(b) and (c) where it is seen that there is a region near the junction where carriers are virtually absent due to recombination. This region is called the depletion layer, and in practice might be about 10^{-4}cm thick.

Although there are very few charge carriers in the depletion layer, the region is actually highly charged [Fig. 13·11(d)] owing to the presence of the ionized impurity atoms. Thus the donor atoms in the *n*-type material are positively charged, while the acceptors on the other side have negative charge. These charges are normally neutralized by the presence of the majority carriers. Since the whole crystal has to be neutral, the numbers of positive and negative charges must be equal. The charge distribution on these impurities is roughly as shown in Fig. 13·11(e).

This charge distribution is rather like that on a parallel-plate capacitor each of whose plates is situated roughly mid-way between the edge of the depletion layer and the junction. The charge per unit area on each 'plate' is equal to the density of donors (or acceptors) multiplied by the width *W* of the depletion

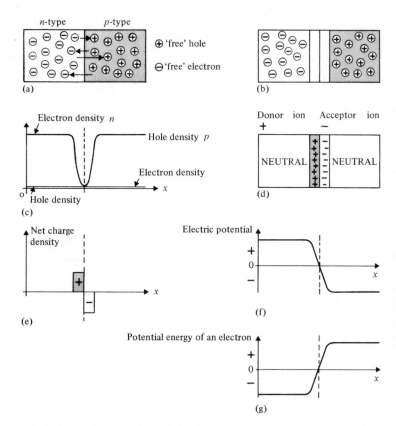

Fig. 13·11 The function of a depletion layer in a *p–n* junction

layer, and it is left as a problem (see problem 13·13) to show that the voltage V_0 across such a capacitor is given approximately by the relation

$$V_0 = \frac{N_d W^2 e}{\varepsilon_r \varepsilon_0}$$

where ε_r is the relative permittivity of the material; in silicon $V_0 = 1·0$ V while in germanium its value is about 0·6 V.

Now this voltage is equal to the work required to be done in moving unit charge across the junction, and the voltage distribution across the junction, shown in Fig. 13·11(f), gives the potential energy distribution of a unit positive charge. It is clear that this distribution prevents the free passage of electrons and holes across the junction without work being done on them. Only a few, those with exceptional amounts of thermal energy, can cross this barrier 'uphill'; the holes in the *p* region are hindered from moving across to the *n* region and similarly motion of electrons towards the *p* region is impeded.

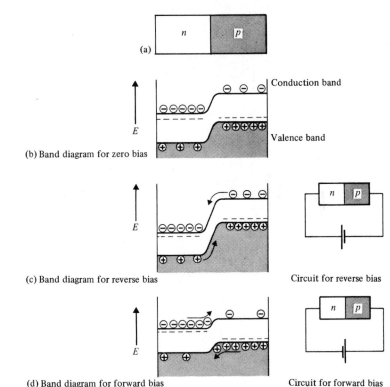

(a)

(b) Band diagram for zero bias

(c) Band diagram for reverse bias

Circuit for reverse bias

(d) Band diagram for forward bias

Circuit for forward bias

Fig. 13·12 The energy band diagram for a biased p–n junction

Such carriers as are able to acquire the extra energy, V_0 electron volts, may cross the barrier, but according to the Boltzmann equation the probability P of an electron having the extra energy V_0 is given by

$$P = \exp\left(-eV_0/kT\right) \tag{13·8}$$

and is a very small probability, as eV_0 is very much greater than kT at room temperature.

Since in equilibrium there can be no net flow of carriers, this small current must be counterbalanced by an equal flow in the opposite direction. Flow in that direction is easy (it is 'downhill' for both positively and negatively charged particles) but there are so few holes on the n side and so few electrons on the p side of the junction that this current is also very small.

Thus the potential gradient and the natural diffusion of carriers act in opposition, setting up a dynamic equilibrium in which no net current flows across the junction.

Since the potential energy of an electron in a potential V is $-eV$, the potential energy of an electron varies across the

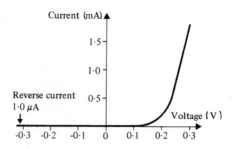

Fig. 13·13 The voltage–current relationship for a silicon *p–n* junction

junction as shown in Fig. 13·11(g). This means that we must redraw the energy band diagrams for electrons in a *p–n* junction as shown in Fig. 13·12(b), where the horizontal axis again represents distance through the crystal, and where the bands are bent to exactly the amount required by the inbuilt potential.

13·10 Current flow through a biased *p–n* junction

The dynamic equilibrium set up by the interdiffusion of holes and electrons is upset when a voltage is applied across the ends of the crystal. To understand the new situation we must first decide what form the energy band diagram takes when the voltage is applied. Let us take the case where the *n*-type region is made electrically positive with respect to the *p*-type region. The first thing to note is that, compared to the bulk of the crystal, the depletion layer has a high resistance because of the absence of free carriers. Thus, remembering that the depletion layer is effectively in series with the rest of the crystal, we see that the applied voltage drop appears almost entirely across the depletion layer. This gives a new energy band diagram as shown in Fig. 13·12(c). The effect of the applied voltage is to increase the potential difference across the junction. This *reduces* the flows of holes into the *n* region and of electrons into the *p* region, while making little difference to the reverse flows since the latter are limited mainly by the scarcity of holes in the *n* region and electrons in the *p* region. The result is a small net current flow from *p* to *n*.

On the other hand, reversing the sign of the bias voltage *V* reduces the height of the potential drop across the junction. Once again the flow of holes and electrons from *n* and *p* regions is scarcely affected. But the flow in the other direction is dramatically increased, because the probability of carriers crossing [see Eqn (13·8)] is increased to

$$P = \exp -e[(V_0 - V)/kT]$$

The probability, and hence the current, is multiplied by the factor $\exp(eV/kT)$, which for $V = 0.1$ V is equal to 52 times!

The current–voltage relationship for an actual silicon p–n junction is shown in Fig. 13·13, where the exponential characteristic is clearly seen. The current in the so called reverse-biased region (V negative in Fig. 13·13) is so small that the junction behaves very nearly as a one way conductor, that is, as a rectifier.

The advantage of semiconductor rectifiers and transistors over their vacuum valve equivalents lies largely in their reliability. There is little that can go wrong with the simple p–n junction and it deteriorates only if so much current is passed that the crystal becomes too hot. When this happens, kT increases and with it the current, causing further heating and eventually destruction of the device.

13·11   Transistors

The transistor is a device which can be used to amplify small alternating voltages or currents. It amplifies by converting power injected from a d.c. source (e.g., a battery) into a.c. power at the frequency of the input signal. In the following simplified description we shall show how power amplification is achieved.

The transistor can be regarded as two p–n junctions connected back-to-back as shown in Fig. 13·14 where the current carriers, holes, and electrons are shown. We now have two potential barriers, one between the (left-hand) *emitter* and (centre) *base* regions and one between the base and the (right-hand) *collector* region. If we apply forward bias, reducing the barrier height between emitter and base regions, holes pass into the base region from the emitter. (We say they are *injected* into the base.)

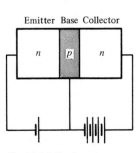

Fig. 13·14 The junction transistor, with its biasing voltages

If the base regions were thick, they would soon recombine with the electrons there. But the base is made especially thin so that, before this can happen, they find themselves swept into the collector, since the potential drop across the base–collector junction encourages this.

Now quite a small forward bias across the emitter–base junction injects a large number of carriers into the base region. At the collector junction a fairly large reverse-bias voltage is applied which assists the flow of holes into the collector from the base while opposing the flow of current in the opposite direction. Thus the current injected from the emitter enters at low voltage, and the same current (if no holes are lost by recombination) is collected at a higher voltage.

If the emitter current is varied in accordance with the magnitude of an a.c. electrical signal, the collector current also varies.

The input power is equal to the a.c. emitter current multiplied by the emitter base voltage, which is small; the output power is greater, being equal to the a.c. emitter current multiplied by the (larger) collector base voltage. The transistor is thus a power amplifier.

Problems

13·1 Using the data in Table 13·1, calculate the mobilities of electrons and holes (assuming them to be equal) in gray tin, silicon, and germanium. Calculate also the mobility which would be necessary for diamond to have an *intrinsic* conductivity of $10^{-12}\Omega^{-1}m^{-1}$ if the hole and electron mobilities were equal.

13·2 How much n-type impurity is necessary to explain the conductivity of $10^{-12}\Omega^{-1}\,m^{-1}$ in diamond if the mobility is $0.18\,m^2/Vs/$electron and intrinsic conductivity can be neglected?

13·3 Calculate the carrier densities in intrinsic silicon and germanium at $100°C$ using the values $N_c = 2.5 \times 10^{25}\,m^{-3}$ and $E_g = 0.75$ eV in germanium and 1.1 eV in silicon. Assuming the hole and electron mobilities to be inversely proportional to temperature, estimate their values at $100°C$ (see Problem 13·1 for the values at $300°K$) and hence calculate the resistivities of silicon and germanium at this temperature.

13·4 Calculate the total number of valence electrons per cubic metre in germanium (density 5·32) and find what fraction of them is available for conduction at $300°K$.

13·5 Calculate the intrinsic conductivity at $300°K$ of the compound semiconductor gallium antimonide (GaSb), for which the energy gap is 0.70 eV. Take $N_c = 2.5 \times 10^{25}\,m^{-3}$ and the electron and hole mobilities to be $2.3\,m^2/v.s$ and $0.010\,m^2/v.s$ respectively.

How much n-type impurity is required to give GaSb a conductivity of $10\,\Omega^{-1}m^{-1}$? Would the same quantity of p-type impurity suffice?

13·6 Since Sb is a Group V element, an excess of Sb in GaSb would make the compound n-type. Using the data given in Problem 13·5, calculate the excess atomic percentage of Sb required to give a conductivity of $10\,\Omega^{-1}\,m^{-1}$. The density of GaSb is $5.4 \times 10^3\,kg/m^3$.

13·7 A piece of p-type silicon contains 10^{24} acceptors per cubic metre. Calculate the temperature at which intrinsic silicon has the same carrier density. Hence deduce qualitatively the manner in which the conductivity of the p-type material depends upon temperature.

13·8 What effect has an increase in temperature on the V–I characteristic of the diode shown in Fig. 13·13? Using Eqn 13·9, calculate the current through the diode under 0.3 V reverse bias at $60°C$, given that Fig. 13·13 is correct at $20°C$.

13·9 How many holes and electrons would you expect to find in a piece of germanium containing an equal number N of donor and acceptor atoms? Calculate the conductivity of a piece of germanium containing 3×10^{22} donors and 8×10^{21} acceptors per cubic metre. The electron mobility in Ge is $0.39\,m^2/v.s$. (Assume all impurity atoms to be ionized and neglect intrinsic carriers.)

13·10 Assuming that in intrinsic silicon at room temperature 10% of the free electrons recombine with holes each second, calculate the approximate rate of release of energy which results. Does this liberation of energy produce a rise in the temperature of silicon?

13·11 In order for an electron to cross from the top of the valence band to the bottom of the conduction band energy must be supplied to it. If the source of energy is a photon calculate the minimum frequency which the photon must have. What bearing does your result have on the optical properties of germanium?

13·12 The resistivity of many polymers (e.g., polyethylene, pvc) is very high. Draw the energy level diagram for electrons in polymers.

13·13 As described in the text, the depletion layers at a *p–n* junction form a kind of capacitor. The capacity may be calculated approximately by assuming all the charge in each layer to be concentrated at the mid-plane of the layer. By writing expressions for the capacity and the charge per unit area of the junction, deduce the expression

$$V_0 = eN_d \, \omega^2 / \varepsilon_r \, \varepsilon_0$$

for the voltage V_0 across the junction. Hence deduce the width of the depletion layer in a silicon *p–n* junction, given $N_d = 10^{24} \, m^{-3}$, $V_0 = 1\cdot0 \, V$, and $\varepsilon_r = 11\cdot7$.

14 Magnetic materials

14.1 Introduction

The ability of certain materials—notably iron, nickel, cobalt, and some of their alloys and compounds—to acquire a large, permanent magnetic moment is of central importance in electrical engineering. The many applications of magnetic properties range from permanent magnets for magnetron oscillators or loudspeakers to soft iron and related materials for transformers and cover the use of almost every aspect of magnetic behaviour.

There is an extremely wide range of differing types of magnetic materials and it is important to know firstly, why these, and only these, materials should display magnetic properties and, secondly, to know what governs the differences between their behaviour. For example, why can one material carry a permanent magnetic moment while another has high permeability?

These two topics will be treated separately in this chapter, beginning with the origin of the magnetic moment.

14.2 Magnetic moment of a body

We know from experiments that if a magnetized solid is subdivided into smaller and smaller parts it is impossible to find a piece which carries a single magnetic charge, or monopole. Magnetic materials act always as dipoles, and for these materials it is possible to define a magnetic dipole moment.

In this respect the material behaves like an assembly of blocks (which we shall later identify as atoms) each carrying a circulating current, i (Fig. 14.1) and having a magnetic dipole moment, ia, where a is the cross-sectional area of the block, that is, the area enclosed by the current loop (see problem 14.1). Thus if the solid is broken down into separate blocks, each block has a dipole moment.

Using this model we can calculate the magnetic moment of the whole body in terms of the dipole moment of each block. Now the net current on the surface of each interior block is zero since neighbouring currents are equal and cancel one another. The only place where no cancellation occurs is at the surface of the body. There is a surface current equal to the current i circulating round each layer of blocks, giving a total current in if there

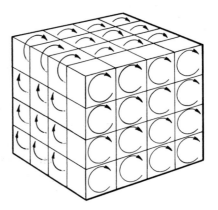

Fig. 14·1 The magnetic moment of a body is the sum of the magnetic moments of the elementary units of which the body is composed

are n blocks along the length of the solid. This current encloses the total cross-sectional area, A, of the body so that the total magnetic moment is $inA = \sum ia$, the sum of the moments of all the blocks.

Clearly the magnetic moment is proportional to volume, so that the magnetic moment per unit volume is a characteristic of the material and not of the shape or size of the solid. The moment per unit volume is called the *intensity of magnetization* (M).

Because we can write magnetic moment in terms of current $\times$ area its units are ampere-metres^2, whence we obtain the units of the intensity of magnetization as amperes per metre.

Now, in the MKS system of units the magnetic flux density, B, through a magnetized body is equal to the intensity of magnetization multiplied by the permeability of free space, μ_0; that is, $B = \mu_0 M$ (no applied field). If an externally applied field, H, is present, we obtain

$$B = \mu_0(H + M) \tag{14·1}$$

So H and M share the same units while B is measured in webers per square metre or teslas. However, μ_0 is usually quoted in henrys per metre so that we have the dimensional relationship:

$$\text{webers/m}^2 = (\text{henrys/m}) \times (\text{amperes/m})$$

Hence,

$$\text{webers} = \text{amperes} \times \text{henrys}$$

We shall meet the relationship between B, H, and M once again, later in this chapter, but in the meantime we return to the problem of the origin of the magnetic moment.

14·3 Atomic magnetic moment

The model of Fig. 14·1 suggests that a solid is divided into the smallest blocks possible (atoms) so that our first task is to consider the magnetic moment of single atoms and later to investigate the magnetic properties of a lattice of atoms.

We begin by considering an isolated atom. As we saw in Chapter 3, the magnetic quantum numbers, m_l and m_s, of an electron are both related to the magnetic moment of that electron. To obtain the total magnetic moment of the atom we have to add up the magnetic moments of the electrons in it taking into account the fact that the direction of the moments is given by the sign of m_l and m_s. Thus, closed shells of electrons having equal numbers of electrons with positive and negative values of m_l and m_s have no net magnetic moment since each electron cancels another. This is shown in Table 14·1 where the dipole moment of each electron is represented by an arrow.

Table 14·1   **Directions of electron moments in a filled 2p shell**

Electron	1	2	3	4	5	6
m_l	-1	-1	0	0	$+1$	$+1$
Moment due to m_l	↓	↓	–	–	↑	↑
m_s	$+\frac{1}{2}$	$-\frac{1}{2}$	$+\frac{1}{2}$	$-\frac{1}{2}$	$+\frac{1}{2}$	$-\frac{1}{2}$
Moment due to m_s	↑	↓	↑	↓	↑	↓

This then is the first important rule to note: the atomic magnetic moment comes only from incomplete electron shells.

Now the valence electrons together form an incomplete shell, so that we would expect all but the rare gases to exhibit magnetic properties. This is true for atoms in isolation, i.e., in the gaseous state, but when we come to consider solids we find only a few elements which are capable of permanently retaining a magnetic moment at room temperature.

If all the elements are studied at very low temperatures a lot more are found to be magnetic while if we include compounds in our investigations yet more materials are found to be strongly magnetic.

However, not just any compound is found to be magnetic, as we show in Table 14·2. In this table the columns and rows represent the groups of the Periodic Table. At each intersection of a row and a column an example is given, where one exists, of a magnetic alloy or compound which is composed of an element from each of the two groups.

Table 14-2 Magnetic materials

Legend:
- *Antiferromagnetic
- †Compound of two Group IIIB elements
- ?Magnetic properties uncertain
- Hyphen denotes a disordered alloy of arbitrary composition

The entries below are arranged by the periodic GROUP of each constituent element. Column headers give the group (and, under VIII, the TRANSITION ELEMENTS Fe, Co, Ni) of one constituent; the left-hand row labels give the group of the other constituent.

Group	IA	IIA	IIIA	IVA	VA	VIA	VIIA	VIII (Fe)	VIII (Co)	VIII (Ni)	IB
IA	—	—	—	—	—	—	—	—	—	—	—
IIA	—	—	—	—	—	—	—	—	—	—	—
IIIA	UH_3*	—	Gd–Sc	—	—	—	—	—	—	—	—
IVA	TiH_2*	—	—	—	—	—	—	—	—	—	—
VA	—	—	—	—	—	—	—	—	—	—	—
VIA	—	—	—	—	CrV*	—	—	—	—	—	—
VIIA	Mn–H	—	YMn_5	—	—	MnCr*	—	—	—	—	—
VIII (Fe)	Fe–H	$FeBe_2$	PrRu	Fe_2Zr	VFe	Fe_3Cr	FeMn	—	—	—	—
VIII (Co)	Co–H	Co–Be	$HoIr_2$	Co_4Zr	Nb–Co	$CrIr_3$	MnRh*	FeRh	—	—	—
VIII (Ni)	Ni–H	Ni_3Mg	$GdPd_2$	Ni–Zr	Ni–V	$CrPt_3$	$MnPt_3$	$FePd_3$	CoPt	—	—
IB	—	—	AgDy*	—	Au_4V	—	$MnAu_4$	FeAu	Co–Au	NiCu	—
IIB	—	—	GdCd	$ZrZn_2$?	—	—	MnHg*	Fe_2Cd	Co–Zn	Ni–Hg_3	—
IIIB	—	—	$CeAl_2$	—	—	$CrAl_2$*	MnB	FeB	Co_2B	Ni_3Al	—
IVB	—	—	$NdGe_2$	—	—	$CrGe_2$	Mn_5Ge_2	FeC	Co–Si	Ni–Si	—
VB	—	—	TbN	—	—	CrAs	MnBi	Fe_2P	Co_5–As_4	Ni–Bi	Cu_2Sb*
VIB	KO_2*	—	EuS	Ti_2O_3*	VSe*	CrTe	Mn_3O_4	Fe_3O_4	CoS_2	Ni_2–Te	CuO*
VIIB	—	—	$GdCl_3$	$TiCl_3$*	VCl_3*	CrI_3	$MnBr_2$*	$FeCl_2$*	CoF_2	NiF_2	AgF_2?
ELEMENTS	—	—	Gd	—	—	Cr*	Mn*	Fe	Co	Ni	—

Additional entry (Group IIIB column): $GaAl_2$†

The table shows that, with one or two exceptions (which can be readily explained) all the magnetic materials contain *at least one transition element.*

The distinguishing feature of a transition element is that one of the inner electron shells is incomplete. This suggests immediately that the inner incomplete shell is the one which gives rise to the atomic magnetic moment, while the valence electrons are of no importance. This characteristic of the valence electrons is just a reflection of the rule which we learnt in Chapter 5 about bonding: atoms bond together in such a way as to form closed shells of electrons. As we saw above, a closed shell carries no magnetic moment.

The exceptions in Table 14·2 may now be explained if we note that the bonding is unusual in these cases: it results in an unfilled inner shell.†

14·4 Size of the atomic magnetic moment

Having established that the magnetic moment of an atom in the solid state is due only to an incomplete inner shell we now calculate what its magnitude should be. In Chapter 3 we noticed that the magnetic moment was related to angular momentum and for each electron this takes two forms: the orbital and the spin angular momenta. For convenience we shall treat these separately. But first we have to discuss the values of m_l and m_s for the electrons in the unfilled shell.

When we considered the filling of energy levels in multi-electron atoms in Chapter 4 no indication was given of the sequence of filling the energy levels in a subshell. The rule governing this was first deduced from the emission spectra of these elements, and is known as Hund's rule after its discoverer. Hund deduced from his spectroscopic measurements that levels of different m_l filled first with electrons having the same m_s value, that is, $m_s = +\frac{1}{2}$. Thus, the first level to fill is that with $m_l = -l$, $m_s = +\frac{1}{2}$; the second level to fill is that with $m_l = -(l-1)$, $m_s = +\frac{1}{2}$, and so on. When, however, the shell is just half full then according to Pauli's principle no more electrons with $m_s = +\frac{1}{2}$ are allowed so that the second half-shell fills in the sequence $m_l = -l$, $m_s = -\frac{1}{2}$ then $m_l = -(l-1)$, $m_s = -\frac{1}{2}$, and so on.

We may summarize this by saying that the electrons arrange themselves among the levels to give the maximum possible total spin angular momentum consistent with Pauli's principle. This is

†For example in cupric oxide (CuO), the copper atom must lose two electrons to form an ionic bond with divalent oxygen and this leaves it with an unfilled $3d$ shell.

the usual form of Hund's rule and we may deduce from it that there must be an interaction between electrons when they have different functions, which tends to make their spin axes parallel. Moreover it seems that two electrons with different wave-functions and with spins parallel have lower energy than two of the same wave-function with antiparallel spins. This cannot be explained without going deeply into the subject of quantum mechanics but we may note that this spin–spin interaction is a direct consequence of Pauli's principle combined with the coulomb repulsion of two like charges.

We can now return to the problem of calculating the atomic magnetic moment. Let us begin with the case of a single orbiting electron. The element scandium is an example of this situation, for there is but one electron in the 3d shell. According to the rules given above, the magnetic quantum numbers of this solitary electron are $m_l = -2$, and $m_s = \frac{1}{2}$. In Chapter 3 it is stated that the magnetic moment is proportional to the angular momentum and we must now derive the exact relationship between them.

We have seen that the magnetic moment, μ, of a current, i, flowing in a loop of radius r is given by

$$\mu = \pi r^2 i \qquad (14\cdot2)$$

In place of i we must put the equivalent current carried by the electron in its orbit. In the present case we wish to calculate the current flow around the circumference of the circular path of radius, $\bar{r}$, the mean radius of the wave-function. The current carried around by the electron can be derived in the manner used earlier to obtain Eqn (12·2). Let us assume that all the electron's charge, $-e$, is moving around at a radius, $\bar{r}$, so that the amount of charge per unit length of the circumference is just $-e/2\pi\bar{r}$. The amount of charge passing a point on the circumference in unit time is then $-ev/2\pi\bar{r}$, where v is the velocity at which the charge is being transported. This velocity is just p/m, that is, the momentum divided by the mass. Thus we can write down the equivalent current, i, as follows

$$i = \frac{-e}{2\pi\bar{r}} \cdot \frac{p}{m} \qquad (14\cdot3)$$

The magnetic moment is obtained by combining Eqn (14·2) and Eqn (14·3):

$$\mu = -\pi\bar{r}^2 \cdot \frac{ep}{2\pi\bar{r}m} = \frac{e}{2m} \cdot p\bar{r} \qquad (14\cdot4)$$

Now the product $p\bar{r}$ is equal to the angular momentum of the electron. Since the total orbital angular momentum is precessing about a fixed axis as described in Chapter 3, only its component along that axis contributes to the atomic magnetic moment. The perpendicular component is continuously rotating (see Fig. 3·10) and averages to zero.

Thus the angular momentum to be inserted in Eqn (14·4) to obtain the orbital magnetic moment μ_{orb} is equal to $m_l h/2\pi$, and the resultant expression for the magnetic moment is

$$\mu_{orb} = -m_l \frac{eh}{4\pi m} \qquad (14·5)$$

Since m_l is an integer, we can regard the quantity $eh/4\pi m$ as a natural unit of magnetic moment. It is called the Bohr magneton, symbol β, and the atom of scandium thus has an orbital magnetic moment of two Bohr magnetons, since its sole $3d$ electron has $m_l = -2$.

The spin magnetic moment is not obtained simply by replacing m_l by m_s in Eqn (14·5) but the relationship is similar

$$\mu_{spin} = -m_s \frac{eh}{2\pi m} \qquad (14·6)$$

so that a single electron with $m_s = \frac{1}{2}$ has a spin magnetic moment equal to one Bohr magneton.

14·5 3d transition elements

The second element in the first transition series is titanium. It has two $3d$ electrons and according to Hund's rule their magnetic quantum numbers are respectively $m_l = -2$, $m_s = \frac{1}{2}$ and $m_l = -1$, $m_s = \frac{1}{2}$.

The total orbital magnetic moment of titanium is the sum of those of the two electrons, that is,

$$\mu_{orb} = \beta + 2\beta = 3\beta \qquad \text{(titanium)}$$

The spin magnetic moment is obtained in the same way but in this case the contributions of the two electrons are both one Bohr magneton:

$$\mu_{spin} = \beta + \beta = 2\beta \qquad \text{(titanium)}$$

In the same way the orbital and spin magnetic moments of vanadium, which has three $3d$ electrons, are both found to be 3β.

So far nothing has been said about the way in which the orbital and spin magnetic moments combine to give the total moment for the atom. This combination is just a little too complicated for consideration here and for our present purpose is unnecessary. For we are confining our attention to the $3d$ series of transition elements, which are different from the $4f$ series because in the former the orbital magnetic moment disappears in the solid state. This is because each atom in the crystal experiences an electric field due to the charge on all the neighbouring ions. This electric field alters the shape of the wavefunctions, and makes the total orbital angular momentum $m_l h/2\pi$ precess in a much more complicated way, with the result that, on the average, there is no component of orbital magnetic moment along any direction. The spin magnetic moment, on the other hand, is not affected by this electric field.

This 'crystal field', as it is called, does not affect the orbital moments of the magnetic electrons in the $4f$ transition metals, starting with lanthanum and cerium, since those electrons are deeper inside the atom and are shielded by the $5d$ and other electrons. Since these shells are full each acts like a charged sphere in shielding anything inside from external electric fields. The conclusion we draw from the above discussion is that the atomic moments of the $3d$ transition elements in the solid state are obtained by adding the spin magnetic moments of the electrons in the unfilled shell.

We have only to note that, at manganese, the $3d$ shell becomes half full, five electrons being present. Because of Pauli's principle the next electron to be added must have $m_s = -\frac{1}{2}$, so that its magnetic moment is opposite to, and subtracts from, those of the first five electrons. Thus iron has a spin moment of 4β, cobalt 3β, and so on. This is shown in Table 14·3.

Table 14·3   Spin distributions and magnetic moments in the first long period

Element:	K	Ca	Sc	Ti	V	Cr	Mn	Fe	Co	Ni	Cu	Zn
Number of $3d$ electrons	0	0	1	2	3	5	5	6	7	8	10	10
Spin directions	—	—	↑	↑↑	↑↑↑	↑↑↑↑↑	↑↑↑↑↑	↑↑↑↑↑ ↓	↑↑↑↑↑ ↓↓	↑↑↑↑↑ ↓↓↓	↑↑↑↑↑ ↓↓↓↓↓	↑↑↑↑↑ ↓↓↓↓↓
Magnetic moment in Bohr magnetons	0	0	1	2	3	5	5	4	4	2	0	0

14·6   Alignment of atomic magnetic moments in a solid

So far we have considered two effects arising from the close proximity of atoms to one another in a solid. These were (1) the forming of closed shells by the valence electrons to give no resultant magnetic moment and (2) the disappearance of the magnetic moment connected with m_l in the case of the unfilled inner $3d$ shell.

We now come to a third effect which has to do with the directions along which the atomic 'magnets' point. In this respect magnetic behaviour may be classified into three basic groups:

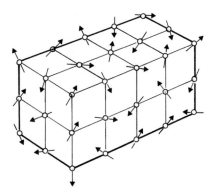

Fig. 14·2 In a paramagnet the atomic magnetic moments (represented by arrows) are all in different directions

(a) Materials with a random distribution of direction of magnetization (Fig. 14·2). An applied field can produce a small degree of alignment but this disappears on removal of the field. This behaviour is termed *paramagnetic*.

(b) Materials in which the magnetic moments on alternate atoms point in opposite directions. As an example Fig. 14·3 shows the arrangement in metallic chromium.

(c) Other solids in which the atomic moments are all parallel (see Fig. 14·4 for iron). This is *ferromagnetism*.

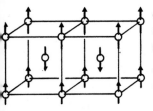

Fig. 14·3 Antiferromagnetic chromium which has the b.c.c. structure

Paramagnetic materials are uninteresting from an engineering viewpoint, so we shall discuss only the last two classes.

14·7 Parallel atomic moments (ferromagnetism)

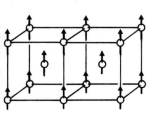

Fig. 14·4 Ferromagnetic iron, which has the b.c.c. structure

The most important of the ferromagnetic materials are all metallic, and, iron, cobalt, nickel, and gadolinium are the only examples which are ferromagnetic at room temperature. The origin of the force which aligns the moments in metals is still a major research problem but it is known to arise from the wave mechanical nature of electrons and it is related to the force which aligns the spins of electrons in a partly filled atomic shell.

There is a further peculiarity about metallic bonding. It is found that in the case of iron, cobalt, and nickel the $3d$ electrons are able to wander through the lattice just like the $4s$ electrons. One effect of this is to reduce the effective magnetic moment of each electron so that instead of the expected values of the atomic magnetic moment calculated above, the measured values are as given overleaf:

Metal	Number of Bohr magnetons per atom (calculated)	Number of Bohr magnetons per atom (measured)
Fe	4	2·22
Co	3	1·72
Ni	2	0·61

If one of these elements is bonded ionically or covalently in a compound, however, the moment is not reduced because none of the 3*d* electrons can move through the lattice.

14·8   Temperature dependence of the magnetization

The magnetic moment of a solid body is the sum of all the individual atomic moments. Thus in metallic nickel, with an atomic moment of $0·61\beta$ and an atomic weight of 58·7, the magnetic moment per kilogramme is

$$\frac{6·02 \times 10^{26} \times 0·61\beta}{58·7} \quad Am^2/kg$$

since by Avogadro's law there are $6·02 \times 10^{26}$ atoms in a kilogramme–molecule.

The magnetic moment per unit volume (called the *intensity of magnetization*, *M*) is the above value multiplied by the density, which for nickel is $8·8 \times 10^3 kg/m^3$. Thus

$$M = \frac{6·02 \times 10^{26} \times 0·61 \times 9·27 \times 10^{-24} \times 8·8 \times 10^3}{58·7}$$

$$= 5·1 \times 10^5 \ A/m$$

Now the measured value of the intensity of magnetization is exactly equal to this at a temperature of 0°K. But at room temperature (293°K) the measurement yields a value of 3·1 A/m, while at 631°K and above nickel is not ferromagnetic! This is explained by the thermal agitation in the lattice which disturbs the alignment of the atomic moments (Fig. 14·5) so reducing the component of each resolved along the direction of the net

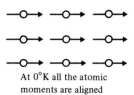

At 0°K all the atomic
moments are aligned

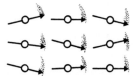

Above 0°K thermal agitation disturbs the alignment
Each moment fluctuates in direction about the average

Fig. 14·5 The effect of thermal agitation on the alignment of atomic moments in a ferromagnet

magnetization. At the *Curie temperature* (631°K for nickel) and above the forces trying to align the moment are not strong enough even partially to overcome the thermal agitation and the atomic moments become randomly oriented—in fact the material behaves like a paramagnet. A graph relating M and T for nickel is shown in Fig. 14·6.

14·9 Antiparallel atomic moments

In chromium metal there are as many moments pointing in one direction as in the opposite direction, resulting in exact cancellation, so that the material has no magnetic moment. This behaviour is called *antiferromagnetic* and it occurs in many compounds as well as some metals. Table 14·2 lists some examples.

There is yet another class of magnetic materials which are like the antiferromagnets in that alternate atoms have magnetic moments pointing in opposite directions. But in this case alternate atoms carry different magnetic moments, as indicated in Fig. 14·7 by the sizes of the arrows. The moments, therefore, do not cancel one another and there is a residual intensity of magnetization. Such materials are termed *ferrimagnetic* and are of great engineering importance since, unlike all the usable ferromagnets they are insulators, being ionic-covalent compounds frequently incorporating oxygen. An example is the oxide of iron, Fe_3O_4, known to the ancients as Lodestone. Here some of the iron atoms are divalent, while the rest are trivalent and hence have a different magnetic moment since one more electron is missing from the $3d$ shell. The net magnetization in this case is due solely to the Fe^{2+} ions since the moments of the Fe^{3+} ions cancel one another.

Further ferrimagnets can be made by substituting other divalent $3d$ transition elements (e.g., Mn, Cr) for the Fe^{2+} ions. These particular magnetic oxides all have the same crystal structure and are called *ferrites*.

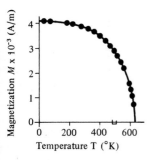

Fig. 14·6 The intensity of magnetization of nickel plotted against the absolute temperature

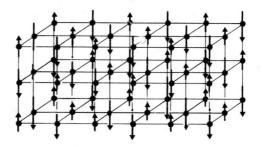

Fig. 14·7 An imaginary ferrimagnet in which alternate magnetic atoms have anti-parallel moments of different magnitudes. In real ferrimagnets the crystal structure is usually more complex

14·10 Magnetization and magnetic domains

In Section 14·5 it was explained that in the ferromagnetic elements the magnetic moments of all the atoms are parallel to one another at 0°K. This would mean that any piece of iron, for example, would be spontaneously magnetized to saturation, that is, it would be a permanent magnet.

On the contrary, in practice a piece of iron is generally unmagnetized in the natural state. On application of a magnetic field, H, it becomes magnetized to an extent determined by its susceptibility to the magnetic field. The intensity of magnetization, M, is related to H by the equation $M = \chi H$, where χ is the magnetic susceptibility. Since M and H share the same units, χ has no dimensions—it is just a number.

The magnetic flux density, B, in a magnetized solid has, as was explained earlier, two components and this is expressed in the relationship

$$B = \mu_0 \, (H + M).$$

Using the relation $M = \chi H$ this can be transformed to read

$$B = \mu_0(H + \chi H) = \mu_0 H(1 + \chi).$$

The proportionality constant $(1 + \chi)$ is often lumped into one symbol: μ_r, the relative permeability; like χ it has no units.

Before going on to consider why M is not just a constant, independent of H, we shall study in more detail the observed behaviour of a piece of iron, for example, when it is placed in a variable magnetic field.

Starting with the iron unmagnetized, we apply an increasing field and measure the flux density produced in the material. Referring to Fig. 14·8, the flux B rises along the line O–P–Q–R, reaching a saturation value, B_{sat}, beyond which any increase in the field has a negligible effect on B.

The intensity of magnetization at this point is equal to the value obtained from a graph like that in Fig. 14·6. If the experiment were performed at 0° K it would also be equal to the value given in Section 14·6 corresponding in iron to $2·22\beta$ per atom.

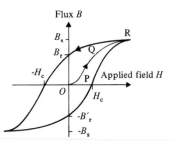

Fig. 14·8 The B–H curve of a ferromagnet or a ferrimagnet

Table 14·4   Properties of some typical magnetic materials at room temperature

Material	Use	B_r (Wb/m²)	B_{sat} (Wb/m²)	H_c (A/m)	μ_r at $H = 0$
Silicon iron (Fe—3% Si)	Transformers	0·8	1·95	24	500 to 1,500
Mu-metal (5Cu, 2Cr, 77Ni, 16Fe)	Magnetic shielding	0·6	0·65	4	2×10^4
Ferrite (48%Mn (48%Mn Fe₂O₄ 52% ZnFe₂O₄)	High frequency transformers	0·14	0·36	50	1,400
Ferrite (NiFe₂O₄)		0·11	0·27	950	15
Carbon steel (0·9 C 1Mm 98Fe)	Permanent magnets	1·0	1·98	4×10^3	14
Alnico V (24Co 14Ni 8Al 3Cu 51Fe)	Permanent magnets	1·31	1·41	$5·3 \times 10^4$	20 to 250
Fe (pure)		1·3	2·16	0·8	10^4
Fe (commerial)		1·3	2·16	7	150

If now the field is reduced again, B does not follow the same curve but decreases more slowly along the line R–B_r; reversing the field and increasing it in the opposite direction causes the flux to follow the line B_r–H_c–S at which point the iron is again saturated but with the magnetization in the opposite direction. Reduction and re-reversal of the field carries the flux along the path S–B'_r–H_c–R and finally closes the loop. The closed path so described is called a *hysteresis loop* and it shows that when the field is removed after reaching the point R, the material retains a remanent flux density, B_r, and behaves as a 'permanent' magnet. In order to destroy the magnetization, it is then necessary to apply a reversed field equal to the coercive force H_c. The value of B_r and H_c may vary widely from material to material and some typical values are shown in Table 14·4.

All the above features can be explained by the following theory.

It was suggested more than fifty years ago that the solid is indeed magnetized to saturation at all times but that it comprises a collection of small volumes called 'domains' each magnetized to saturation but each with its magnetization in a direction which is random with respect to all the others. The collection of randomly oriented domains gives the solid a zero overall magnetic moment. The boundaries between domains are called domain walls and subsequent research has led to ways of making these walls visible, confirming the theory. It has also been shown that in many cases the domains are not random but quite regular. As might be expected the domains in a single

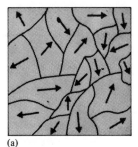

(a)

Fig. 14·9 (a) Hypo-
thetical domain
arrangement in a
demagnetized
ferromagnet
(b) Domains in a
polycrystalline sample
of nickel

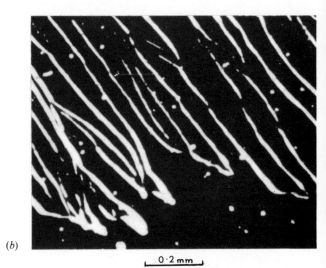

(*b*)

$\llcorner$ 0·2 mm $\lrcorner$

crystal are more regular than those in a polycrystalline solid. We show examples in Figs. 14·9 and 14·10.

The magnetization curve and the hysteresis loop can be interpreted in terms of domains. As an example, we take a single crystal in which the process is clear. On first application of the field the domain walls move so as to increase the volume of those domains whose direction of magnetization is nearest to the field direction [Fig. 14·10(b)]. As the unfavourably oriented domains shrink they become more resistant to entire removal, and rotation of magnetization within whole domains begins to play a larger part in the process. In some soft magnetic materials (those with high permeability) rotation of the magnetization is very easy, and occurs at much lower applied fields simultaneously with domain wall motion.

Saturation is only complete when all the unfavourable domains are removed—this occurs only at very high fields, of several times the nominal saturation field as determined from the *B–H* loop.

This, then, explains qualitatively the shape of the initial magnetization curve (O–P–Q–R in Fig. 14·8) but does not explain why in most solids the same curve is not followed on reduction of the field. In the case of the single crystal in Fig. 14·10 the initial magnetization curve is followed quite closely in the reverse direction but most practical materials contain large numbers of imperfections which act as obstacles to rotation of the magnetization and domain wall motion. When the field is increased these obstacles are overcome by the energy supplied by the field but when the field is removed the defects prevent the domain walls from returning to their previous position. The situation is rather like that of a horse pulling a cart along a

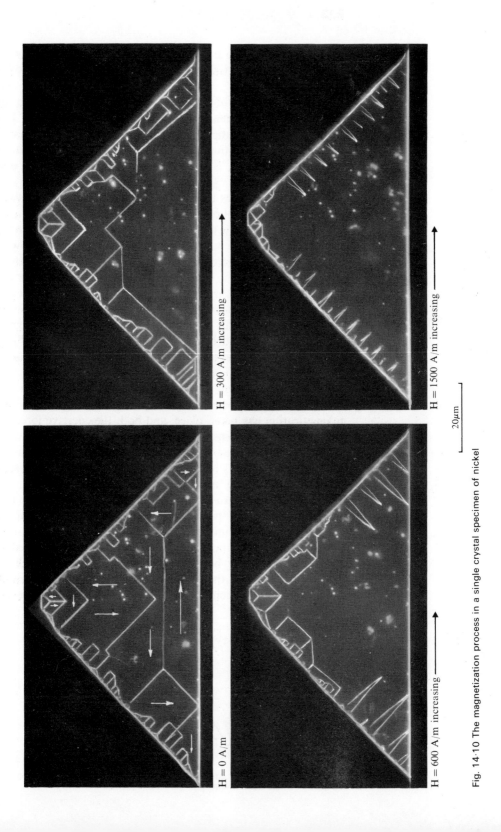

H = 0 A/m

H = 300 A/m increasing

H = 600 A/m increasing

H = 1500 A/m increasing

20µm

Fig. 14·10 The magnetization process in a single crystal specimen of nickel

hilly road: when the horse is unhitched the cart does not roll back right to its starting point but merely to the bottom of the last hill.

Thus, on removal of the field the flux does not return to zero but to a remanent magnetization which can be quite a large fraction of the saturation value. To return the domain structure to a random distribution with zero net moment it is necessary to supply more energy, either from a reversed applied field or in the form of heat.

This model enables us to understand the difference between hard and soft magnetic materials in a qualitative way. In a soft material there are few obstacles to wall motion or rotation; the coercive force is low and the magnetization changes follow the applied field, as it increases or decreases, with little hysteresis.

From Table 14·4 it can be seen that careful purification of iron both increases the relative permeability and decreases the coercive force compared to iron of a commercial grade. Silicon iron, which contains 3% silicon, is another soft magnetic material which is much used in transformers. It has a slightly lower hysteresis loss than iron and also lower eddy current losses owing to its higher resistivity.

Conversely, if many obstacles to domain wall motion exist the domains will lock in the fully magnetized position after magnetization has taken place, and both the coercive force and the remanent flux density will be large. Such a material would make a good permanent magnet and some permanent magnets are made from alloys in which a precipitate forms, the precipitated particles acting as obstacles.

Yet another modern kind of permanent magnet is made from collections of small particles whose size is smaller than the thickness of a domain wall. They are thus unable to contain more than one domain. They are made with an elongated shape (the aspect ratio is between 5:1 and 10:1) which makes it difficult to reverse their magnetization. They are normally magnetized along their length (see Fig. 14·11) and to reverse

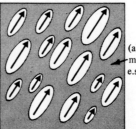

(a) Non-magnetic material between e.s.d. particles

Fig. 14·11 A permanent magnet may be made of a collection of elongated single domain particles; the magnetization directions are shown

them the magnetization must rotate (no walls being present) and hence must be made to pass through the state of magnetization perpendicular to their axis. This requires a very large applied field since the problem is like that of trying to magnetize a needle transverse to its length.

Collections of such particles in a non-magnetic bonding medium make good permanent magnets because, being single domains, the remanence of each particle is high and hence so is that of the whole assembly. In addition, the large field required for reversal results in a high coercivity. The hysteresis loop of the material is thus very open, almost square. Such materials are used in the manufacture of magnetic recording tape, which consists of elongated particles of ferric oxide (Fe_2O_3).

Problems

14·1 Calculate the couple on a circular current loop of area a carrying a current i when placed in a magnetic flux B which makes an angle θ with the axis of the loop. What is the corresponding couple on a short bar magnet whose magnetic moment (i.e., pole strength × distance between poles) is m? Hence deduce the magnetic moment of the current loop.

14·2 What are the magnetic moments, in Bohr magnetons, of the following ions? $Ru^+, Nb^+, La^{2+}, Gd^{3+}, Pt^{2+}, Fe^{2+}, Fe^{3+}$?

14·3 Calculate the intensity of magnetization at $0°K$ of iron, in which the atomic magnetic moment is $2·22$ Bohr magnetons and the density is $7·87 \times 10^3$ kg/m^3.

14·4 The chemical formula for Fe_3O_4 can be expressed in the form $(Fe^{2+}O)(Fe_2^{3+}O_3)$, so that there are twice as many Fe^{3+} ions as Fe^{2+} ions. Calculate the intensity of magnetization of Fe_3O_4 at $0°K$, given that the density is $5·18 \times 10^3$ kg/m^3.

14·5 In manganese ferrite, $MnFe_2O_4$, the Mn^{2+} ions take the place of the Fe^{2+} ions in Fe_3O_4, while the dimensions of the unit cell are almost identical. Calculate the intensity of magnetization at $0°K$.

14·6 The magnetic moment of a paramagnetic material is directly proportional to the applied magnetic field. Aluminium is paramagnetic and has a susceptibility of $1·9$ per cubic metre. Calculate the field required to give aluminium the same intensity of magnetization as nickel at room temperature (see Fig. 14·7). Is it possible to attain such a field in the laboratory?

14·7 A toroid of ferrite has the following dimensions: internal diameter $1·0$ cm, outside diameter $2·0$ cm, thickness $0·5$ cm. Calculate the total flux through the specimen when it is magnetized circumferentially by a field of 100 A/m, if the intensity of magnetization is $12,000$ A/m.

14·8 Why does Mumetal (see Table 14·4) make a good magnetic shield? What other material in the table would also serve? Why is it not used in practice?

14·9 What effect should cold working have on the properties of Mumetal? How would you restore its original properties after fabrication of a shield?

14·10 The coercivity of a material is raised by incorporating microscopic voids which impede the motion of domain walls. What other magnetic property is influenced by the voids?

14·11 According to the discussion in Chapter 3 of the gyroscopic nature of magnetic moments, one would expect that, when a magnetic field is applied to a magnetized body, its magnetic moment should precess around the field. Why does this not occur?

15 Dielectrics

15·1 Introduction

Dielectric materials or insulators have the unique property of being able to store electrostatic charge. Very often the material is charged up by friction as in the classic school experiment of rubbing a glass rod with dry silk. Virtually all the modern plastic materials are good dielectrics: the charging up of nylon fabric through friction with other clothing gives rise to crackling sparks and sometimes to tangible arcs and shocks when the garment is removed in a dry atmosphere. One of the authors has seen a three-inch spark from a nylon garment in the dry atmosphere of the South African high veldt. Bearing in mind that an arc struck in dry air at 5,000 feet altitude needs a field of approximately 10^7 volts per metre, such a spark corresponds to a potential of 400,000 volts! Before attempting to account for this phenomenon we must first consider the dielectric in terms of the ideas of energy bands discussed in connection with semiconductors in Chapter 13.

15·2 Energy bands

Dielectric materials are invariably substances in which the electrons are localized in the process of bonding the atoms together. Thus covalent or ionic bonds, or a mixture of both, or van der Waals bonding between closed-shell atoms all give rise to solids (or gases) exhibiting dielectric (insulating) properties. The energy band diagram for a crystal will be just like that of a semiconductor with a valence band and a conduction band separated by an energy gap as was mentioned in Chapter 13. The gap is so large that, at ordinary temperatures, thermal energy is insufficient to raise electrons from the valence to the conduction band, which is, therefore, empty of electrons. Consequently, there are no free charge carriers and the application of an electric field will produce no current through the material.

This description applies to a perfect dielectric: in practice there will always be a few free electrons in the conduction band. These will be knocked there by stray high-energy radiation (such as cosmic rays) and irradiation by visible or ultra-violet light and will have a relatively long 'lifetime' before returning to the valence band since, once the electrons are free, the

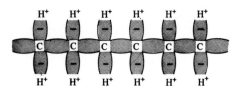

Fig. 15·1 Paraffin molecule showing polar side arms

probability of their being recaptured by an empty bond is low because so few bonds are empty. However, when the energy gap exceeds about three electron volts the number of such electrons is so small that they are unable to give a significant current. In a good insulator the current, when a potential of several hundred volts is applied, will be of the order of 10^{-9}A or less.

It should be remembered that the energy band concept is strictly relevant only to a single crystal of material. This is because it assumes that every atom and its bonding system is the same as every other so that an electron requires precisely the same energy to be liberated from a bond anywhere in the solid. Many dielectrics used as insulators are highly disordered so that the environment of each atom tends to be a little different from that of its neighbours, as in the glassy network structure described in Chapter 10. However, it is still possible to consider the energy band picture to apply but with the band edges smeared out somewhat. This allows for the fact that slightly different energies may be needed to energize an electron from a bond at different places in the material.

In the electrostatic charging phenomenon mentioned in the introduction charge is stored on the surface of the material where it persists because there are no free carriers to neutralize it, and the charge itself becomes bound in the surface. Various mechanisms can be postulated whereby charge may be bound in the surface of the material but none is easy to demonstrate experimentally. In a material with a repetitive although not necessarily regular structure like glass, the surface must represent an interruption of the bonding system. There will be atoms which have been unable to complete a covalent or ionic bond with a neighbour because there is no neighbour with which to do so; there will be one or two valence electrons which are relatively loosely bound to their parent atoms and these may be transferred into the material used for the rubbing by mechanical work due to friction. The surface will then be left with a positive charge due to the loss of the negatively charged electrons. It is a demonstrable fact that a glass rod acquires positive charge on being rubbed. The positive charge will remain so long as it is unable to acquire replacement negative charge to compensate it. When there is water vapour in the air the H_2O molecules will readily ionize to OH^- and H^+ and the hydroxyl (OH^-) groups become attached to the surface to replace the lost negative charge. Thus the charge only persists in a dry atmosphere. In practice, in the process of fabrication of glass the surface usually becomes covered with hydroxyl ions and the frictional work tends to remove these rather than to remove electrons, as described above. This results in a positively charged surface just the same.

The case of polymer-based materials, such as resin, bakelite, silk, nylon, and so on, must be somewhat different. This is evidenced by the fact that, on rubbing they acquire a negative charge. The structure of such materials comprises covalently bonded, long–chain molecules held together by van der Waals forces and there are no loosely–bound electrons in the surface. However, all these molecules have side-groups which are like electrostatic dipoles. The simplest case, that of the paraffins, is illustrated in Fig. 15·1.

Since the bonding electrons are concentrated mainly midway between the carbon and hydrogen ions the positive hydrogen ion represents a positive charge spatially displaced from the negative charge in the bond, so forming a dipole. This occurs with other kinds of side-groups so that water molecules, which are dipolar, can attach themselves by forming hydrogen bonds with the side-groups. The action of the mechanical work due to friction may then be to ionize the water, that is, to 'wipe off' a hydrogen ion, leaving the negatively charged hydroxyl group attached to the end of the polymer side chain. This would then give a negative surface charge. Again, in a damp atmosphere the hydroxyl could readily capture a hydrogen ion and so neutralize the surface charge.

It is emphasized that the above mechanisms are only tentative explanations and that the actual processes involved in the frictional surface charging of dielectrics are not yet well understood. However, the general features of the explanations are well founded and lead to the conclusion that, since the sites for charge absorption are atoms or groups of atoms, it will be possible for a very large density of surface charge to be accumulated. For example, it can be shown that there are 10^{11} to 10^{12} 'dangling bonds' on one square centimetre of the surface of a covalently bonded insulator. Thus the maximum charge density could be in the region of 10^{12} electrons per square centimetre, that is, $1·6 \times 10^{-19} \times 10^{12} = 1·6 \times 10^{-7}$ coulombs per square centimetre. In order to estimate what this implies in terms of voltages we shall first recapitulate the basic laws of electrostatics.

15·3 Coulomb's law

Experiments on electrically charged bodies yield the following observations:

1. Like charges repel and opposite charges attract each other.
2. The force between charges is:
   (a) inversely proportional to the square of the distance between them;
   (b) dependent on the medium in which they are embedded;
   (c) acts along the line joining the charges;
   (d) is proportional to the product of the charge magnitudes.

These facts are summarized in Coulomb's law of force which gives the force, **F**, between charges as

$$\mathbf{F} = \frac{q_1 q_2}{4 \pi \varepsilon r^2} \mathbf{a} \text{ newtons} \tag{15·1}$$

where q_1 and q_2 are the charges in coulombs, r is the distance between them in metres, and ε is called the permittivity of the medium in which they are embedded. The vector **a** is a unit vector pointing along the line joining the charges and reflects the fact that when the charges have the same sign the force is one of repulsion. The units of permittivity may be deduced from Eqn (15·1) thus:

$$\varepsilon = \frac{(\text{coulombs})^2}{\text{newtons} \ (\text{metres})^2}$$

The properties of a material as a dielectric enter into Coulomb's law only through this permittivity, which is also a measure of its ability to store charge. This follows from Coulomb's law of capacitance which states that the capacity of a body to store charge is defined by the equation

$$Q = CV \tag{15·2}$$

where $+Q$ is the charge on one surface of the body and $-Q$ is the charge on the opposite surface, V is the potential drop between the surfaces, and C is the capacitance of the body. The dimensions of C are given by Eqn (15·2) as

$$\text{Capacitance} = \frac{\text{coulombs}}{\text{volts}}$$

and the unit of capacitance is called the farad. The capacitance of a body is found experimentally to be an intrinsic property of the material which forms the body and of its geometry, that is, its physical shape. It is most conveniently defined in terms of a parallel plate capacitor. If we have two metal electrodes, each of area a square metres, separated by a distance,

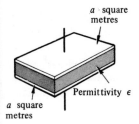

a · square metres

a square metres

Permittivity ε

Fig. 15·2 Parallel-plate capacitor

d metres and parallel to each other, filled with a material of permittivity, ε, (Fig. 15·2), the capacitance of the system is given by

$$C = \frac{\varepsilon a}{d} \qquad (15\cdot3)$$

It can be shown that the potential energy stored by a capacitor is given by

$$E = \frac{1}{2}\frac{Q^2}{C} = \tfrac{1}{2}CV^2 \qquad \text{joules} \qquad (15\cdot4)$$

when it has Q coulombs of charge stored. From this we see that

$$\text{joules} = \text{newton metres} = \frac{(\text{coulombs})^2}{\text{farads}}$$

Combining this with Eqn (15·1) shows that the dimensions of permittivity reduce to farads per metre. When the medium in the capacitor is empty space, the permittivity is written as ε_0 and has the value $\frac{1}{36\pi} \times 10^{-9}$ farads per metre.

The permittivity, ε, of a dielectric material may be related to the permittivity of vacuum by writing it as

$$\varepsilon = \varepsilon_r \varepsilon_0 \qquad (15\cdot5)$$

where ε_r is the relative permittivity and is simply a number. It has exactly the same value as the dielectric constant, k, defined in the CGS system of units. Values of relative permittivity for a

Table 15·1   Relative permittivity and loss factors of various materials

Material	Relative Permittivity (ε) at 1 MHz	Loss Factor (tan δ) at 1 MHz
Alumina	4·5–8·4	0·0002–0·01
Amber	2·65	0·015
Glass (Pyrex)	3·8–6·0	0·008–0·019
Mica	2·5–7·0	0·0001
Neoprene (rubber)	4·1	0·04
Nylon	3·4–3·5	0·03–0·04
Paraffin wax	2·1–2·5	0·003 ($\approx$ 900Hz)
Polyethylene	2·25–2·3	0·0002–0·0005
Polystyrene	2·4–2·75	0·0001–0·001
Porcelain	6·0–8·0	0·003–0·02
PTFE (Teflon)	2·0	0·0002
PVC (rigid)	3·0–3·1	0·018–0·019
PVC (flexible)	4·02	0·1
Rubber (natural)	2·0–3·5	0·003–0·008
Titanium dioxide	14–110	0·0002–0·005
Titanates (Ba, Sr, Ca, Mg, and Pb)	15–12,000	0·0001–0·02

range of materials are given in Table 15·1. It should be noted that dry air has a value approximately equal to one, the relative permittivity of vacuum.

We may now return to the question of the potential of a friction-charged surface. We may take the surface as one plate of a capacitor with the other being the nearest earthed surface, that is, the nearest surface at zero potential.

Let us take the example of the three-inch spark quoted in the introduction to the chapter. The cross-sectional area of the arc is small, say one square millimetre, and so only that tiny area of the nylon surface is discharged by the spark. We can calculate the charge stored on that area, knowing that the potential across the three-inch capacitor is 400,000 volts, and from the capacity, C, which is given by,

$$C = \frac{\varepsilon_0 a}{d}$$

where $a = 1$ mm^2 and $d = 0.076$m. Thus

$$C = 4\pi \times 10^{-7} \times 10^{-6}/0.076 = 1.7 \times 10^{-11} \text{ farads.}$$

The charge stored is then

$$Q = CV = 1.7 \times 10^{-11} \times 4 \times 10^5 = 6.8 \times 10^{-7} \text{ coulombs}$$

This charge would correspond to Q/e single electronic charges, that is to $(6.8 \times 10^{-7})/(1.6 \times 10^{-19}) = 4.2 \times 10^{12}$ electronic charges in an area of one square millimetre. This is an electron charge density of 4.2×10^{14} per square centimetre. Now, in a Nylon fabric there are about 10^{15} dipolar side-chains per square centimetre of surface so that this rough calculation suggests that on average each side-chain acquires a single electronic charge as a result of friction.

15·4   Complex permittivity

Permittivity, $\varepsilon = \varepsilon_r \varepsilon_0$, has been defined as a property of the medium by means of Coulomb's law. If a parallel plate capacitor has a capacitance, C_0, in air, and the space between the plates is then filled by a medium of relative permittivity ε_r then, neglecting fringing effects, the capacitance becomes $C = \varepsilon_r C_0$.

When an alternating electromotive force, V, is applied across an ideal capacitor an alternating current, i, will flow, its value being determined by the reactance of the capacitor, the value of which is given by $1/\omega C$, where $\omega = 2\pi f$, f being the frequency. Now a fundamental property of a capacitor is that the current that flows to and from it due to its charging and discharging successively is 90° out of phase with the alternating voltage

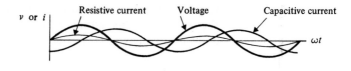

Fig. 15·3 Voltage–current phase relationships in a capacitor

applied across it. This means that the current and voltage are related as shown in Fig. 15·3. In real capacitors, containing a dielectric, the phase angle is found not to be exactly 90° but is less than 90° by some small angle, δ. This is due to the flow of a small component of current which is in phase with the applied voltage as the current would be in a resistor. This resistive current shown in Fig. 15·3, combines with the capacitive current to give a total current which is $(90-\delta)°$ out of phase with the applied voltage. The resistive current is due to the dielectric actually acting as a very poor conductor and it is often referred to as leakage current. The magnitude of the leakage current is a property of the dielectric and can be represented in a circuit by showing the capacitor to have a resistance in parallel with it, as in Fig. 15·4. The value of the resistance is determined by the size of the leakage current.

To allow for this phenomenon mathematically we represent the relative permittivity by a complex number so that

$$\varepsilon_r = \varepsilon' - j\varepsilon'' \tag{15·6}$$

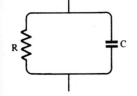

Fig. 15·4 Equivalent circuit of lossy dielectric

Here ε' represents the part of relative permittivity that increases capacitance and ε'' represents the 'leakage' or 'loss'. The derivation of this expression is given, for those familiar with j-notation in a.c. theory, in an extension at the end of the chapter. It is shown there that the 'loss angle', δ, is given by

$$\tan \delta = \frac{\varepsilon''}{\varepsilon'} \tag{15·7}$$

and tan δ is called the loss factor of the dielectric. This can range in value from 10^{-5} for the very best insulators to 0·1 for rather poor dielectrics. Some typical loss factors are included in Table 15·1.

15·5   Electric flux density

When a dielectric material is placed in an electric field already existing in a homogeneous medium such as air, it has the effect of changing the distribution of the field to a degree depending upon its relative permittivity, that is, the electric field intensity is a function of the medium in which it exists. We represent this situation mathematically by defining an electric flux density **D**, in the dielectric by the equation

$$\mathbf{D} = \varepsilon \mathbf{E} \qquad (15\cdot8)$$

where electric field **E** is a vector, since it has magnitude and direction, and so therefore is the flux density. The units of flux density may be derived from Eqn (15·8) and are coulombs per square metre. Now, electric field at a point is defined as the force per unit charge on a positive test charge placed at that point. Thus, from Eqn (15·1),

$$\mathbf{E} = \frac{q}{4\pi\varepsilon r^2}\,\mathbf{a} \qquad (15\cdot9)$$

will be the field at a point distant r from a charge q.

Combining Eqns (15·9) and (15·8) we have

$$\mathbf{D} = \frac{q}{4\pi r^2}\,\mathbf{a} \qquad (15\cdot10)$$

and we see that, since ε does not appear in the equation, the flux density arising from a point charge, q, is independent of the medium and is a function of the charge and its position only. If we take q to be at the centre of a sphere of radius, r, the flux density will be the same at all points on the surface. The total flux, $\varPsi$, crossing the surface will be the flux density multiplied by the area of the sphere, hence, using Eqn (15·10),

$$\varPsi = 4\pi r^2\,\frac{q}{4\pi r^2} = q \text{ coulombs}$$

Thus we see that the total flux crossing the surface of a sphere with a charge at its centre is equal to the value of the charge.

This statement is generalized in Gauss' law which states that, for any closed surface containing a system of charges, the flux out of the surface is equal to the charge enclosed. Mathematically this is written as

$$\oint D \cos \alpha\, dS = \sum q \qquad (15\cdot11)$$

where α is the angle between the normal to the surface and D, the flux vector; dS is an element of surface area, as shown in Fig. 15·5. The integral represents the total normal component of flux over the closed surface. We shall be using this idea later in the chapter. In vector notation Eqn (15·11) is written as

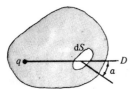

Fig. 15·5 Illustration of Gauss' Law

$$\oint \mathbf{D}.d\mathbf{S} = \sum q \qquad (15\cdot12)$$

15·6   Polarization

When connected to a battery an air capacitor will charge until the free charges on each plate produce a potential difference equal and opposite to the battery voltage. A dielectric increases the charge storage capacity of a capacitor by neutralizing some of the free charges which would otherwise contribute to the potential difference opposing the battery voltage. More charge can, as a result, flow into the capacitor which then has an increased storage capacity given by $\varepsilon_r C_0$, where C_0 is the original capacity in air. We visualize this effect as arising from alignment of electrostatic dipoles in the dielectric under the influence of the field between the capacitor plates. The dipoles form long chains with a positive charge at one end and a negative charge at the other. The positive charge will be adjacent to the negative capacitor plate and will neutralize some of the charge on it. Similarly the negative end of the dipole chain will neutralize some of the charge on the positive capacitor plate.

For an applied battery voltage V, the charge carried by the air capacitor will be $Q_0 = C_0 V$ and, on insertion of the dielectric, $Q = \varepsilon_r C_0 V$.

Since we are not, in this argument, considering loss mechanisms, ε_r may be replaced by ε' (ε'' assumed zero) and we may write

$$\frac{Q}{\varepsilon'} = C_0 V \quad \text{or} \quad V \propto \frac{Q}{\varepsilon'}$$

The implication is that, of the total charge Q, only a fraction, Q/ε', contributes to neutralization of the applied voltage, the remainder $Q[1-(1/\varepsilon')]$ being bound charge neutralized by the polarization of the dielectric. We define the polarization of the dielectric in terms of this bound charge. The polarization, P, is equal to the bound charge per unit area of the dielectric surface and is measured in coulombs per square metre—the same units as the flux density, **D**; like **D**, the polarization is a vector quantity.

Thus we may imagine the electric flux density in a dielectric to be due to two causes: the flux density, which would be set up in the space occupied by the dielectric by an applied electric field, and the polarization of the dielectric which results from the electric field. Thus we may write,

$$\mathbf{D} = \varepsilon_0 \mathbf{E} + \mathbf{P} \tag{15·13}$$

In the example of the parallel plate capacitor discussed above, **D**, **E**, and **P** are all parallel to each other so that the magnitudes D, E, and P may be used in Eqn (15·13), that is, $D = \varepsilon_0 E + P$.

In Eqn (15·8) we defined D such that

$$\mathbf{D} = \varepsilon' \varepsilon_0 \mathbf{E} \qquad (15\cdot14)$$

and using this with equation (15·13) we have

$$\mathbf{P} = \mathbf{D} - \varepsilon_0 \mathbf{E} = \varepsilon_0 \mathbf{E}(\varepsilon' - 1) \quad \text{and} \quad \mathbf{P} = \chi \mathbf{E} \qquad (15\cdot15)$$

where $\chi = \varepsilon_0 (\varepsilon' - 1)$ is called the dielectric susceptibility of the medium and is given by

$$\chi = \frac{\text{bound charge density}}{\text{free charge density}}$$

The measurement of polarization (which is the analogue of magnetization of a magnetic material) is based on Eqn (15·15) and consists in the measurement of χ or, in practice, of ε'. This is dealt with at the end of the chapter.

15·7 Mechanisms of polarization

Permittivity is essentially a macroscopic, or averaged out, description of the properties of a dielectric. To understand exactly what is happening in the material when an electric field is applied, we have to link the permittivity to atomic or molecular mechanisms which describe the processes of polarization of the material.

On the macroscopic scale we have defined the polarization, **P**, to represent the bound charges at the surface of the material.

Two point electric charges, of opposite polarity, $+Q$ and $-Q$, separated by a distance, d, represent a dipole of moment, μ, given by

$$\mu = Qd \qquad (15\cdot16)$$

The moment is a vector whose direction is taken to be from the negative to the positive charge.

We now have, in the polarized dielectric, P bound charges per unit area and if we take unit areas on opposite faces of a cube separated by a distance l, the moment due to unit area will be

$$\mu = Pl \qquad (15\cdot17)$$

For unit distance between the unit areas $l = 1$ and we have $\mu = P$ per unit volume. Thus the polarization, P, is identical with the electric moment per unit volume of the material. This moment may be thought of as resulting from the additive action of N elementary dipoles per unit volume, each of average moment $\bar{\mu}$, therefore

$$P = N\bar{\mu} \qquad (15\cdot18)$$

Furthermore, μ may be assumed to be proportional to a local electric field inside the dielectric. If this is denoted by E_{int}, being the value of the field acting on the dipole, we define

$$\mu = \alpha E_{int} \qquad (15 \cdot 19)$$

where α is called the *polarizability* of the dipole, that is, the average dipole moment per unit field strength. The dimensions of α are

$$\frac{(\text{second})^2 \, (\text{coulomb})^2}{\text{kilogrammes}} = \varepsilon \times (\text{metre})^3$$

Thus we have

$$\mathbf{P} = (\varepsilon - 1) \, \varepsilon_0 \, \mathbf{E} = N \alpha \, \mathbf{E}_{int} \qquad (15 \cdot 20)$$

This is referred to as the Clausius equation.

Since α is defined in terms of dipole moment, its magnitude will clearly be a measure of the extent to which electric dipoles are formed by the atoms and molecules. These may arise through a variety of mechanisms, any or all of which contribute to the value of α. Thus, for convenience, we regard the total polarizability to be the sum of individual polarizabilities each arising from one particular mechanism, i.e.,

$$\alpha = \alpha_e + \alpha_a + \alpha_d + \alpha_i$$

where the terms on the right-hand side are the individual polarizabilities which will now be discussed.

15·8 Optical polarizability (α_e)

An atom comprises a positively charged inner shell surrounded by electron clouds having symmetries determined by their quantum states. When a field is applied the electron clouds are displaced slightly with respect to the positive cores causing the atoms to take on an induced dipole moment. This induced moment has all the characteristics of an assembly of dipoles produced by elastic displacement of electrons. The strength of the induced moment, μ_e, for an atom is proportional to the local field in the region of the atom and is given by Eqn (15·19) as

$$\mu_e = \alpha_e E_{int}$$

where α_e is called the optical polarizability; it is sometimes also referred to as the electronic polarizability. This is, of course, the mechanism discussed in connection with van der Waals bonding.

15·9 Molecular polarizability (α_a and α_d)

Consider a diatomic molecule made up of atoms A and B. Because of the interaction between them there is a redistribution of electrons between the constituent atoms which should, generally, be axially symmetrical along AB. It may be expected that the diatomic molecule will possess a dipole moment in the direction AB, except where the atoms A and B are identical, when the dipole moment should vanish for reasons of symmetry. Molecules having a large dipole moment are described as 'polar', an example being hydrochloric acid in which there is a displacement of charge in the bonding between H^+ and the Cl^- ions. This gives rise to a configuration in which a positive charge is separated from a negative charge by a small distance, thus forming a true dipole.

Under the influence of an applied field the polarization of a polar substance will change by virtue of two possible mechanisms. Firstly, the field may cause the atoms to be displaced, altering the distance between them and hence changing the dipole moment of the molecule. This mechanism is called atomic polarizability, α_a. Secondly, the molecule as a whole may rotate about its axis of symmetry, so that the dipole aligns itself with the field. This is referred to as *orientational polarizability* (α_d).

15·10 Interfacial polarizability (α_i)

In a real crystal there inevitably exists a large number of defects such as lattice vacancies, impurity centres, dislocations, and so on. Free charge carriers, migrating through the crystal under the influence of an applied field may be trapped by, or pile up against, a defect. The effect of this will be the creation of a localized accumulation of charge which will induce its image charge on an electrode and give rise to a dipole moment. This constitutes a separate mechanism of polarization in the crystal, and is given the name *interfacial polarizability* (α_i).

Any or all of the above mechanisms may contribute to the behaviour in an applied field. As has been described earlier, they are lumped together in a phenomenological constant, α, the polarizability, defined by Eqn (15·19).

15·11 Classification of dielectrics

In general any or all of the above mechanisms of polarization may be operative in any material. The question is how can we tell which are the important ones in a given dielectric? We can do this by studying the frequency dependence of permittivity.

Imagine, first of all, a single electric dipole in an electric field.

It will, given time, line itself up with the field so that its axis lies parallel with the field; if the field is reversed, the dipole will turn itself round through 180° so that it again lies parallel with the field. When the electric field is an alternating one the dipole will be continually switching its position in sympathy with the field. For an assembly of dipoles in a dielectric the same will apply, the polarization alternating in sympathy with the applied field. If the frequency of the field increases a point will be reached when, because of their inertia, the dipoles cannot keep up with the field and the alternation of the polarization will lag behind the field. This corresponds to a reduction in the apparent polarization produced by the field, which appears in measurements as an apparent reduction in the permittivity of the material. Ultimately, as the field frequency increases, the dipoles will barely have started to move before the field reverses, and they try to move the other way. At this stage the field is producing virtually no polarization of the dielectric. This process is generally called *relaxation* and the frequency beyond which the polarization no longer follows the field is called the relaxation frequency.

Considering now the various mechanisms of polarization we can predict, in a general way, what the relaxation frequency for each one might be.

Electronic polarizability relies on the position of electrons relative to the core of an atom. Since the electrons have extremely small mass they have little inertia and can follow alternations of the electric field up to very high frequencies. In fact, relaxation of electronic polarizability is not observed until the visible or ultra-violet light range of the frequency spectrum.

In atomic polarizability we require individual ions to change their relative positions. Now we know that these atoms vibrate with thermal energy and the frequencies of the vibrations correspond to those of the infra-red wavelengths of light. Thus the relaxation frequencies are in the infra-red range.

Orientational polarizability refers to actual reorientation of groups of ions forming dipoles. The inertia of these groups may be considerable so that relaxation frequencies may be expected to occur in the radio frequency spectrum.

In the case of interfacial polarizability a whole body of charge has to be moved through a resistive material and this can be a very slow process. The relaxation frequencies for this mechanism can be as low as fractions of a cycle per second.

In Fig. 15·6 we show a curve of the variation of relative permittivity, ε', with the logarithm of frequency over the entire spectrum, as it might be expected to occur for a hypothetical material showing all these effects.

The basic electronic and atomic polarizabilities, α_e, α_a, and α_d, lead to a general classification of dielectrics. All dielectrics fall into one of three groups:

(a) non-polar materials which show variations of permittivity in the optical range of frequencies only; this includes all those dielectrics having a single type of atom, whether they be solids, liquids, or gases;

(b) polar materials having variation of permittivity in the infra-red as well as the optical region; the most important members of this group are the ionic solids, such as rock-salt and alkali halide crystals in general;

(c) dipolar materials which, in addition, show orientational polarization; this embraces all materials having dipolar molecules of which one important common one is water. The chemical groups O—H and C=O are dipolar and may impart dipolar properties to any material in which they occur.

15·12 Piezoelectricity

'Piezo' is derived from the Greek word meaning 'to press', and the piezoelectric effect is the production of electricity by pressure. It occurs only in insulating materials and is mani-fested by the appearance of charges on the surfaces of a single crystal which is being mechanically deformed. It is easy to see the nature of the basic molecular mechanism involved. The application of stress has the effect of separating the centre of gravity of the positive charges from the centre of gravity of the negative charges, producing a dipole moment. Clearly, whether or not the effect occurs depends upon the symmetry óf the distributions of the positive and negative ions. This restricts the effect so that it can occur only in those crystals not having a centre of symmetry since, for a centro-symmetric crystal, no combination of uniform stresses will produce the necessary separation of the centres of gravity of the charges. This descrip-tion makes it clear that the converse piezoelectric effect must exist. When an electric field is applied to a piezoelectric crystal it will strain mechanically. There is a one-to-one correspondence between the piezoelectric effect and its converse, in that crystals for which strain produces an electric field, will strain when an electric field is applied.

The *pyroelectric effect*, as the name implies, is connected with heat. Certain crystals, such as tourmaline, acquire an electric charge when heated; they are termed pyroelectric. The fact that all pyroelectric crystals are piezoelectric as well leads to some difficulties—when heated, the crystal will expand and

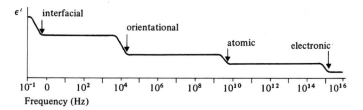

Fig. 15·6 Frequency spectrum of relative permittivity for hypothetical material

may be deformed. Because of the piezoelectric effect, the crystal will acquire an electric charge by virtue of the deformation: this may be termed the *indirect* pyroelectric effect. There is also a direct effect in which there is a change of positive and negative polarization on certain portions of the crystal due simply to the change of temperature produced by uniform heating. In experimental observations the direct and indirect effects are difficult to separate, and there would be little point in so doing since pyroelectric behaviour can be considered simply as a feature of piezoelectric crystals and related to piezoelectricity, because all pyroelectric crystals are also piezoelectric.

Crystals can be divided into 32 classes on the basis of their symmetry. Of these, 20 possess the property of piezoelectricity by virtue of their low symmetry and of these 20, ten are pyro-electric.

Piezoelectric crystals have been widely used to control the frequency of electronic oscillators. If the crystal is cut in the form of a thin slab it will have a sharp mechanical resonance frequency and, in a suitable circuit, this resonance can be excited by an applied alternating voltage, the frequency of which it then controls. Quartz has been widely used for this purpose as it has reasonably high piezoelectric activity. We can define piezoelectric activity in various ways; one way is to define a piezoelectric coefficient, d, by

$$d = \frac{\partial S}{\partial E}$$

where S is the strain and E the field producing it. For quartz the value of d varies with crystallographic direction and has a maximum value 2.25×10^{-12} coulombs/newton or metres/volt.

Another important field of application of piezoelectric materials is as ultrasonic transducers. These are devices which, when excited by an alternating electric field, vibrate mechanically and set up sound waves. Conversely, sound falling upon a suitably designed transducer will cause it to generate alternating voltages which may be amplified, in other words it is a microphone. Many modern record player pick-ups employ

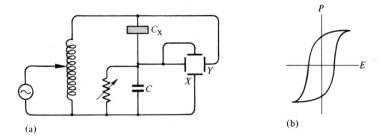

Fig. 15·7 (a) Circuit for a ferroelectric hysteresis loop tracer; (b) Hysteresis loop for BaTiO$_3$

piezoelectric materials to convert mechanical vibration from the needle to electrical signals.

Much development has taken place in these materials in the last two decades. Ammonium dihydrogen phosphate (NH$_4$H$_2$PO$_4$), known as ADP is piezoelectric and can be produced as large crystals. It has d-coefficients in the region of 5×10^{-11} coulombs/newton. Similar values are obtainable with EDT (C$_6$H$_{14}$N$_2$O$_6$) and DKT (K$_2$C$_4$H$_4$O$_6$·$\frac{1}{2}$H$_2$O) which also show very small mechanical losses when vibrating. A most effective and fairly new material, which is made in ceramic form, is lead zirconate titanate (PZT) which has a d-coefficient in the region of 10^{-10} coulombs/newton.

15·13   Ferroelectricity

Reference has been made to the ten crystal classes exhibiting pyroelectricity. This is the name given to the spontaneous polarization, present in polar crystals, which is a function of temperature.

Ferroelectric materials are a subgroup of the pyroelectrics. In the straightforward pyroelectric the direction of spontaneous polarization cannot be reversed by the application of an electric field; in the ferroelectric it can. Thus the ferroelectric crystal can be 'switched' by application of a field and a hysteresis loop is associated with the switching. If a graph of applied field against polarization is drawn a typical hysteresis loop is obtained, similar to the $B–H$ loop for a ferromagnetic.

A ferroelectric hysteresis loop may readily be demonstrated on an oscilloscope by means of the circuit in Fig. 15·7(a) and a typical loop for BaTiO$_3$ is shown in Fig. 15·7(b). In the absence of switching, the voltages across C and C_x will be inversely proportional to their capacities and, neglecting losses, will be in phase. The oscilloscope screen shows a straight line at a small angle to the horizontal if C has a fairly large capacitance. When the ferroelectric crystal, C_x, switches there is a transfer of the surface counter-charge, Q_s, round the circuit such that if, in the initial state, there was a charge $+Q_s$ on the upper face of the crystal there will be $-Q_s$ after switching. The total charge transfer is therefore $2Q_s = 2P_sA$, where P_s is the spontaneous polarization and A is the electrode area. The hysteresis effect arises because a threshold value of the field is needed across the crystal before it switches. When switching occurs the voltage across C will change by $2Q_s/C$, giving a Y-deflection proportional to the spontaneous polarization, P_s, of the crystal, provided that all the ferroelectric domains are switched. Domains are regions in which the polarization is all in a given direction and in most crystals there is a large number of such regions in which the polarization has different orientations. The reversal of the net macroscopic dipole moment of the crystal occurs through nucleation and growth of domains favourably oriented with respect to the applied field.

The existence of a hysteresis loop is proof of the presence of spontaneous polarization in the crystal; the absence of a loop, however, is not proof that there is no spontaneous polarization, but simply that it cannot be reversed by the field applied.

For a given material, ferroelectricity is generally exhibited below a specific temperature, called the Curie temperature. Above this temperature, thermal agitation is sufficient to destroy the cooperative ordering of the dipoles which gives rise to the spontaneous polarization. Thus the transition from a para-electric to a ferroelectric state is essentially a disorder-to-order change.

These are essentially phase changes in the thermodynamic sense discussed in Chapter 9. Ferroelectric materials can be classified as uniaxial, which can polarize only 'up' or 'down' along one crystal direction, or multiaxial, which can polarize along several axes that are equivalent in the non-polar state. Some of the better known ferroelectrics are given below.

Uniaxial ferroelectrics
Rochelle Salt and related tartrates
Potassium dihydrogen phosphate and related materials (XH_2YO_4, with $X = K$, Rb or Cs and $Y = P$ or As)
$(NH_4)_2SO_4$ and $(NH_4)_2BeF_4$
Colemanite, CaB_3O_4 (OH_3) . H_2O
Thiourea, $(NH_2)_2CS$
TGS (triglycine sulphate) $(CH_2NH_2COOH)_3H_2SO_4$ and the related salts such as triglycine fluoberyllate (TGF)
Guanidine aluminium sulphate hexahydrate (GASH) (CN_3H_6) $Al(SO_4)_26H_2O$
Lithium selenite $LiH_3(SeO_3)_2$ and related salts.

Multiaxial ferroelectrics

Barium titanate ($BaTiO_3$) and related materials with perovskite crystal structure

Certain niobates of the type of lead metaniobate $Pb(NbO_3)_2$ with a pyrochlore type of crystal structure

Certain alums of the type $XY(SO_4)_2 12H_2O$, where X is an ammonium ion and Y a specially symmetrical ion like Al or Cr

Ammonium cadmium sulphate $(NH_4)_2Cd_2(SO_4)_3$

The above is not a complete list of known ferroelectrics, since new ones are being discovered continuously.

The electrical analogues of antiferromagnetism and ferrimagnetism are also found. Lead zirconate ($PbZrO_3$) is ferroelectric in high fields, but not in low. X-ray analysis has shown that, at room temperature, there is an antiparallel array of ionic displacements in the material, giving it an antiferroelectric character. The oxide $NaNbO_3$ has been shown to have antiparallel electric dipoles of unequal moments which classifies the material as ferrielectric.

15·14 Molecular mechanisms

A large number of possible mechanisms of ferroelectricity has been discussed and, to date, no completely satisfactory quantitative theory has been forthcoming. It is clear from careful X-ray analysis, however, that the dipole moment is associated with distortion of molecular groups, while the cooperative alignment of these dipoles, to give spontaneous polarization, is a function of the interatomic bonding in the crystal.

In the case of the uniaxial crystals, and of sulphates in the multiaxial group, the dipole moment is due to the deformation of atomic groups such as SO_4, SeO_4, AsO_4, etc., in which the undeformed state is a symmetrical arrangement of the oxygen ions around the central sulphur (or other) ion. At the Curie temperature the crystal strains spontaneously, the central ion is slightly displaced, and the atomic groups acquire a dipole moment. The spontaneous strain is due to ordering of the hydrogen bonds (present in all the uniaxial ferroelectrics), corresponding to the order–disorder phase change in the crystal. The ordered bonds act to align the induced dipoles causing spontaneous polarization. The determining factor in the order–disorder phase change is the (thermodynamic) free energy associated with the crystal structure, and is outside the scope of the present text.

The case of the multiaxial ferroelectrics of perovskite and pyrochlore structure differs somewhat. Here the basic molecular structure is an octahedral arrangement of oxygen ions around a central ion such as Ti in $BaTiO_3$. Below the Curie point, the Ti ion is shifted from its symmetrical position, giving a dipole moment to the molecular group. The alignment of the moments, however, is not due to hydrogen bonds but to coupling between the oxygen ions. This coupling is partially ionic and partially covalent, and the directional property of the covalent bond is responsible for the alignment.

15·15 Dielectric behaviour

The total permittivity of normal dielectrics decreases with decreasing temperature. In ferroelectric materials the permittivity and susceptibility increase with decreasing temperature, going through a sharp maximum at the Curie temperature and thereafter falling further, as temperature decreases. Above the Curie temperature the dielectric susceptibility follows a Curie–Weiss law of a type which is also encountered in magnetism, that is,

$$\chi \doteqdot \frac{C}{T-T_c} \qquad (15·27)$$

where C is called the Curie constant. For ferroelectrics containing hydrogen bonding, C is of the order of $100°K$. The temperature dependence of permittivity and of spontaneous polarization of potassium dihydrogen phosphate (KDP) are shown in Fig. 15·8(a) and (b) and for Rochelle salt in Fig. 15·9(a) and (b) over the region of their Curie temperatures.

The high permittivity, typically of the order of 10,000 at the

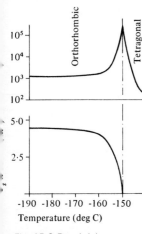

Fig. 15·8 Permittivity and polarization of KDP as a function of temperature

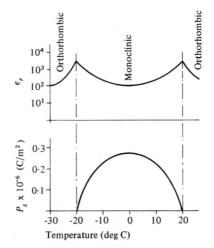

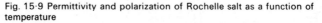

Fig. 15·9 Permittivity and polarization of Rochelle salt as a function of temperature

Curie temperature, is readily understandable. At this temperature the ions are on the point of moving into or out of the position corresponding to spontaneous polarization; consequently an applied field will be able to produce relatively large shifts, with big changes in dipole moment, corresponding to a high permittivity. The falling susceptibility below T_c corresponds to an increasing degree of spontaneous saturation of polarization in the material.

The high relative permittivities of ferroelectric materials are potentially of considerable practical importance, and much effort has been put into taking advantage of them in capacitors. Piezoelectric behaviour and chemical instability as well as the rapid variation of permittivity with temperature have, however, limited their application.

*15·16 Derivation of complex permittivity

We assume a capacitor of value C_0 in air to be filled with a medium of relative permittivity ε_r. When an alternating voltage V is applied across it the current, i, will be 90° out of phase with the voltage and will be given by

$$i = j\omega \varepsilon_r C_0 V \tag{15·28}$$

provided that the dielectric is a 'perfect' one. This current is just the normal capacitor charging and discharging current, which is 90° out of phase with the applied voltage and consumes no power. In general, however, an in-phase component of current will appear, corresponding to a resistive current between the capacitor plates. Such current is due entirely to the dielectric medium and is a property of it. We therefore characterize it as a component of permittivity by defining relative permittivity as

$$\varepsilon_r = \varepsilon' - j\varepsilon'' \tag{15·29}$$

Combining this with Eqn (15·28) we have

$$i = j\omega(\varepsilon' - j\varepsilon'') C_0 V \quad \text{or}$$
$$i = \omega\varepsilon'' C_0 V + j\omega\varepsilon' \tag{15·30}$$

Thus the current has an in-phase component $\omega\varepsilon'' C_0 V$ corresponding to the observed resistive current flow. The magnitude of ε'' will be defined by the magnitude of this current.

It is conventional to describe the performance of a dielectric in a capacitor in terms of its *loss angle*, δ, which is the phase angle between the total current, i, and the purely quadrature component i_c.

If the in-phase component is i_L, then

$$|i| = (|i_L|^2 + |i_c|^2)^{1/2} \tag{15·31}$$

and

$$\tan \delta = \left| \frac{i_L}{i_c} \right| = \frac{\omega \varepsilon'' C_0 V}{\omega \varepsilon' C_0 V} = \frac{\varepsilon''}{\varepsilon'} \tag{15.32}$$

This is the result quoted as Eqn (15·7).

*15·17   Measurement of permittivity

A wide variety of methods exists for the measurement of permittivity, the particular method adopted being determined by the nature of the specimen and the frequency range in which the measurement is to be made.

The measurement of the real part of the relative permittivity, ε', is generally done by measuring the change in capacitance of a capacitor, brought about by the introduction of the dielectric between its electrodes. The imaginary part, ε'', is found by measurement of tan δ, the loss factor resulting from the introduction of the dielectric.

The most usual type of measurement employs a bridge circuit working in the range 10^2 to 10^7 Hz. The commonest circuit is the Schering bridge illustrated in Fig. 15·10. In this, C represents the dielectric-filled capacitor and R the dielectric loss. It should be noted that, as shown in Fig. 15·4, the loss is represented by a resistance in *parallel* with the capacitor. However, this can always be simulated by a series resistor R_s which will produce the same relationship between current and voltage as a parallel resistor R_p in parallel with a capacitor C_p. The relationships between the components are then

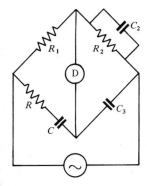

$$R_p = \frac{1}{\omega^2 C_s^2 R_s} \quad \text{and} \quad R_s = \frac{1}{\omega^2 C_p^2 R_p}$$

In the bridge circuit, C_2 and C_3 are standard calibrated capacitors and the resistors R_1 and R_2 are usually made equal. In use, C_2 and C_3 are adjusted until there is zero current through the detector D.

Initially, without the dielectric present, $C = C_0$ and $R = 0$; if, in addition, $R_1 = R_2$, the balance equations are

$$C_0 = C_3 \quad \text{and} \quad C_2 = 0$$

With the dielectric inserted, the equations become

$$C = C_3 \quad \text{and} \quad R = \frac{R_1 C_2}{C_3} \tag{15.33}$$

Since $\tan \delta = \omega C R$ for a series R–C circuit, we have $\tan \delta = \omega C_2 R_1$ and, since R_1 is fixed, the dial of C_2 may be calibrated directly in values of tan δ. Using $\varepsilon''/\varepsilon' = \tan \delta$ and $C/C_0 = \varepsilon'$, the value of the complex permittivity for the dielectric is found.

Problems

15·1 State Coulomb's law of force between electric charges. A positive sodium ion and a negative chlorine ion are a distance 5 Å apart in vacuum. Calculate the attractive force between them if each has a charge of magnitude equal to the charge on the electron. What will be the value of this force if the ions are embedded in water of relative permittivity 80? What is the relevance of your calculation to the solubility of rock salt in water?

15·2 A flat plate of ebonite of area 10 cm^2 is rubbed so that its surface is charged electrostatically. If the charge is equivalent to 10^{14} electrons per square centimetre, what will be the potential difference between the ebonite and a metal plate of the same area placed in air a distance 1 cm from it? At what distance would we expect a spark to jump from the ebonite to the plate if the breakdown strength of the air is 10^9 V/m?

15·3 Explain why a complex permittivity is attributed to a lossy dielectric material. Show that the flux density due to a point charge is independent of the medium in which it is embedded.

In a parallel plate capacitor the distance between the plates, each of area 1 cm^2 is 1 mm. The dielectric has a relative permittivity of 10 and it holds 10^{-9} coulombs of charge. What is the flux density in the dielectric?

15·4 Summarize the atomic and molecular mechanisms of polarization indicating, with reasons, in what range of frequency you would expect relaxation for each one.

15·5 Explain what is meant by the term ferroelectric when applied to a dielectric material. How would you determine whether or not a material was: (a) piezoelectric, (b) pyroelectric, (c) ferroelectric?

Further reading

General
The following books cover most of the material in this volume, at a greater depth (chapter numbers are given in brackets):

Wulff, J., *et al. The structure and properties of materials* (4 vols), John Wiley, New York, 1965.

Wert, C. A. and R. M. Thomson, *Physics of solids*, McGraw–Hill, New York, 1964.

Structure of matter
Moffatt, W. G., G. W. Pearsall, and J. Wulff, *The structure and properties of materials* (Vol. 1 Structure), John Wiley, New York, 1965. (1–6, 11)†

Evans, R. C., *An introduction to crystal chemistry* (2nd edn), Cambridge University Press, London, 1964. (5, 6)

Cottrell, A. H., *Theory of crystal dislocations*, Blackie, London, 1964. (7)

Hume-Rothery, W., *Atomic theory for students of metallurgy*, Institute of Metals, London, 1946. (1–5, 10)

O'Driscoll, K. F., *The nature and chemistry of high polymers*, Chapman and Hall, London, 1965. (11)

Mechanical properties
Cottrell, A. H., *The mechanical properties of matter*, John Wiley, New York, 1964. (8–10)

Hayden, H. W., W. G. Moffatt, and J. Wulff, *The structure and properties of matter*, (Vol. 3 Mechanical properties), John Wiley, New York, 1965. (7–9,15)

Sharp, H. J.(ed), *Engineering materials*, Heywood, London, 1966. (7–12)

Kelly, A., *Strong solids*, Oxford University Press, London, 1966. (8)

Thermodynamics
Sears, F. W., *An introduction to thermodynamics, the kinetic theory of gases and statistical mechanics*, Addison-Wesley, Reading, Mass., 1950. (10)

Brophy, J. H., R. M. Rose, and J. Wulff, *The structure and properties of matter* (Vol. 2 Thermodynamics of structure), John Wiley, New York, 1965. (9, 10)

† Numbers in parentheses indicate the corresponding chapters in this book.

Electrical properties

Pease, L., R. M. Rose, and J. Wulff, *The structure and properties of matter* (Vol. 4 Electronic properties), John Wiley, New York, 1965. (12–15)

Azaroff, L. V., and J. J. Brophy, *Electronic processes in materials*, John Wiley, New York, 1963. (12–15)

Hemenway, C. L., M. Coulter, and R. W. Henry, *Physical electronics* (2nd edn), John Wiley, New York, 1967. (1, 2, 12,13)

Data books

Hodgman, C. D., (ed), *Handbook of chemistry and physics*, Chemical Rubber Publishing Company, New York. (Published annually.)

Anderson, J. C., *et al.*, *Data and formulae for engineering students*, Pergamon Press, Oxford, 1967.

Kaye, G. W. C. and T. H. Laby, *Table of physical and chemical constants*, Longmans Green, London, 1959.

Appendix 1
Units and conversion factors

SI units

(1) *Basic units*

quantity	unit	unit symbol
length	metre	m
mass	kilogramme	kg
time	second	s
electric current	ampere	A
thermodynamic temperature	degree Kelvin	°K
luminous intensity	candela	cd

(2) *Derived units*

quantity	unit	unit symbol
force	newton	$N = kg\,m/s^2$
work, energy, heat	joule	$J = N\,m$
power	watt	$W = J/s$

electrical units

potential	volt	$V = W/A$
resistance	ohm	$\Omega = V/A$
charge	coulomb	$C = A\,s$
capacitance	farad	$F = A\,s/V$
electric field strength	—	V/m
electric flux density	—	C/m^2

magnetic units

magnetic flux	weber	$Wb = V\,s$
inductance	henry	$H = V\,s/A$
magnetic field strength	—	A/m
intensity of magnetization	—	A/m
magnetic flux density	—	Wb/m^2

(3) *Unit conversion factors*

Length, volume

1 mil   = $2 \cdot 54 \times 10^{-5}$ m
1 in   = 2·54 cm   1 in^3 = 16·39 cm^3
1 ft   = 0·3048 m   1 ft^3 = 0·02832 m^3
1 mile = 5,280 ft = 1·609 km
1 μm = 10^{-6} m = 39·37 μin
1 Å   = 10^{-10} m
1 gal   = 0·1605 ft^3 = 4,546 cm^3
1 US gal = 0·1337 ft^3 = 3,785 cm^3

Mass

1 lb   = 0·4536 kg
1 ton = 2,240 lb = 1,016 kg

Density
$$1 \ lb/in^3 = 27 \cdot 68 \ g/cm^3$$
$$1 \ lb/ft^3 = 16 \cdot 02 \ kg/m^3$$
$$1 \ g/cm^3 = 10^3 \ kg/m^3$$

Force
$$1 \ pdl = 0 \cdot 1383 \ N$$
$$1 \ lbf = 32 \cdot 17 \ pdl = 4 \cdot 448 \ N$$
$$1 \ tonf = 9{,}964 \ N$$
$$1 \ kgf = 2 \cdot 205 \ lbf = 9 \cdot 807 \ N$$
$$1 \ dyn = 10^{-5} \ N$$

Power
$$1 \ hp = 550 \ ft \ lbf/s = 0 \cdot 7457 \ kW$$
$$1 \ ft \ lbf/s = 1 \cdot 356 \ W$$

Torque
$$1 \ lbf \ ft = 1 \cdot 356 \ N \, m$$
$$1 \ tonf \ ft = 3{,}037 \ N \, m$$

Energy, work, heat
$$1 \ ft \ lbf = 1 \cdot 356 \ J$$
$$1 \ kWh = 3 \cdot 6 \ MJ$$
$$1 \ Btu = 1{,}055 \ J = 252 \ cal = 778 \cdot 2 \ ft \ lbf$$
$$1 \ cal = 4 \cdot 187 \ J$$
$$1 \ hp \ h = 2 \cdot 685 \ MJ$$
$$1 \ erg = 10^{-7} \ N \, m$$

Pressure, stress
$$1 \ lbf/in^2 = 0 \cdot 07031 \ kgf/cm^2 = 6{,}895 \ N/m^2$$
$$1 \ tonf/in^2 = 157 \cdot 5 \ kgf/cm^2 = 15 \cdot 44 \ N/mm^2$$
$$1 \ kgf/cm^2 = 0 \cdot 09807 \ N/mm^2$$
$$1 \ lbf/ft^2 = 47 \cdot 88 \ N/m^2$$
$$1 \ ft \ H_2O = 62 \cdot 43 \ lbf/ft^2 = 2{,}989 \ N/m^2$$
$$1 \ in \ Hg = 70 \cdot 73 \ lbf/ft^2 = 3{,}386 \ N/m^2$$
$$1 \ mm \ Hg = 1 \ torr = 133 \cdot 3 \ N/m^2$$
$$1 \ Int \ atm = 1 \cdot 013 \times 10^5 \ N/m^2 = 14 \cdot 70 \ lbf/in^2$$
$$1 \ bar = 10^5 \ N/m^2 = 14 \cdot 50 \ lbf/in^2$$

Temperature
$$1^\circ C = 1 \cdot 8^\circ F$$
$$T^\circ K = T^\circ C + 273 \cdot 15^\circ C$$
$$T^\circ R = T^\circ F + 459 \cdot 67^\circ K$$

Dynamic viscosity
$$1 \ poise \ (g/cm \ s) = 0 \cdot 1 \ kg/ \ m \ s = 0 \cdot 1 \ Ns/m^2$$
$$1 \ kgf \ s/m^2 = 0 \cdot 9807 \ kg/m \ s$$
$$1 \ lb/ft \ h = 0 \cdot 4134 \ g/m \ s$$
$$1 \ lbf \ s/in^2 = 6{,}895 \ kg/m \ s$$

Kinematic viscosity
$$1 \ ft^2/s = 0 \cdot 09290 \ m^2/s$$
$$1 \ in^2/s = 6 \cdot 452 \ cm^2/s$$

Thermal conductivity
   1 Btu/ft h deg R = 1·731 J/m s deg K
   1 cal/cm s deg K = 418·7 J/m s deg K

Electrical units

The conversion factors which follow are from the CGS system to the SI system. (Note: in the CGS system 1 e.m.u. = 10^{10} e.s.u. of charge.)

capacitance	1 e.s.u.	$= \frac{1}{9} \times 10^{-11}$ F
charge	1 e.m.u.	$= 10$ C
current	1 e.m.u.	$= 10$ A
electric field strength	1 e.s.u.	$= 3 \times 10^{4}$ V/m
electric flux density	1 e.s.u.	$= (1/12\pi) \times 10^{-5}$ C/m^2
electric polarization	1 e.s.u.	$= \frac{1}{3} \times 10^{-5}$ C/m^2
inductance	1 e.m.u.	$= 10^{-9}$ H
intensity of magnetization	1 e.m.u.	$= 10^{3}$ A/m
magnetic field strength	1 e.m.u.	$= (1/4\pi) \times 10^{3}$ A/m
magnetic flux	1 e.m.u.	$= 10^{-8}$ Wb
magnetic flux density	1 e.m.u.	$= 10^{-4}$ Wb/m^2
magnetic moment	1 e.m.u.	$= 10^{-3}$ A m^2
magnetomotive force	1 e.m.u.	$= (10/4\pi)$ A
mass susceptibility	1 e.m.u/g	$= 4\pi \times 10^{-3}$ /kg
potential	1 e.m.u.	$= 10^{-8}$ V
resistance	1 e.m.u.	$= 10^{-9}$ Ω

Appendix 2
Physical constants

Avogadro's number	N	$= 6 \cdot 023 \times 10^{26}/\text{kg mole}$
Bohr magneton	β	$= 9 \cdot 27 \times 10^{-24} \text{ A m}^2$
Boltzmann's constant	k	$= 1 \cdot 380 \times 10^{-23} \text{ J/deg K}$
electron volt	eV	$= 1 \cdot 602 \times 10^{-19} \text{ J}$
electronic charge	e	$= 1 \cdot 602 \times 10^{-19} \text{ C}$
electronic rest mass	m_e	$= 9 \cdot 109 \times 10^{-31} \text{ kg}$
electronic charge to mass ratio	e/m_e	$= 1 \cdot 759 \times 10^{11} \text{ C/kg}$
energy for T = 290°K	kT	$= 4 \times 10^{-21} \text{ J}$
energy of ground state H atom (Rydberg energy)		$= 13 \cdot 60 \text{ eV}$
Faraday constant	F	$= 9 \cdot 65 \times 10^{7} \text{ C/kg mole}$
permeability of free space	μ_o	$= 4\pi \times 10^{-7} \text{ H/m}$
permittivity of free space	ε_o	$(1/36\pi) \times 10^{-9} \text{ F/m}$
Planck's constant	h	$= 6 \cdot 626 \times 10^{-34} \text{ J s}$
proton mass	m_p	$= 1 \cdot 672 \times 10^{-27} \text{ kg}$
proton to electron mass ratio	$m_p \, m_e$	$= 1,836 \cdot 1$
Average radius of wave-function of H atom in ground state		$= 0 \cdot 529 \times 10^{-10} \text{ m}$
		$= 0 \cdot 529 \text{ Å}$
standard gravitational acceleration	g	$= 9 \cdot 807 \text{ m/s}^2$
		$= 32 \cdot 17 \text{ ft/s}^2$
Stefan–Boltzmann constant	σ	$= 5 \cdot 67 \times 10^{-8} \text{ J/m}^2 \text{ s deg K}^4$
		$= 4 \cdot 76 \times 10^{-13} \text{ Btu/ft}^2 \text{ s deg R}^4$
universal constant of gravitation	G	$= 6 \cdot 67 \times 10^{-11} \text{ N m}^2/\text{kg}^2$
		$= 3 \cdot 32 \times 10^{-11} \text{ lbf ft}^2/\text{lb}^2$
universal gas constant	R	$= 8 \cdot 314 \text{ kJ/kg mole deg K}$
		$= 1,545 \text{ ft lbf/lb mole deg R}$
velocity of light in vacuo	c	$= 2 \cdot 9979 \times 10^{8} \text{ m/s}$
volume of 1 kg mole of ideal gas at N.T.P.		$= 22 \cdot 42 \text{ m}^3$

Appendix 3
Physical properties of elements

Atomic number	Element	Density at 20°C (10^3 kg/m³)	Melting point (°C)	Boiling point (°C)	Young's modulus (10^{10} N/m²)	Shear modulus (10^{10} N/m²)	Poisson's ratio
1	Hydrogen (H)	0·00009	−259·2	−252·7			
2	Helium (He)	0·00018		−268·9			
3	Lithium (Li)	0·534	180·5	1,330	1·15		
4	Beryllium (Be)	1·848	1,277	2,770	29·65	14·48	0·08
5	Boron (B)	2·34	(2,100)	(2,550)			
6	Carbon (C)						
	Graphite	2·25	3,700ᵃ	4,830	0·69 (Polycrystalline)		
	Diamond	3·52			82·74	34·48	0·25
7	Nitrogen (N)	0·00125	−210	−196			
8	Oxygen (O)	0·00143	−219	−183			
9	Fluorine (F)	0·0017	−219·6	−188			
10	Neon (Ne)	0·0009	−249	−246			
11	Sodium (Na)	0·971	97·8	892	0·9		
12	Magnesium (Mg)	1·74	650	1,105	1·72		0·3
13	Aluminium (Al)	2·699	660	2,450	6·9	2·62	0·34
14	Silicon (Si)	2·33	1,410	2,680	11·0		
15	Phosphorous (P) (white)	1·83	44·2	280			
16	Sulphur (S) (yellow)	2·07	119	445			
17	Chlorine (Cl)	0·0032	−101	−34·7			
18	Argon (A)	0·0018	−189·4	−186			
19	Potassium (K)	0·86	63·7	760	0·34		
20	Calcium (Ca)	1·55	838	1,440	2·07	0·69	0·31
21	Scandium (Sc)	2·99	1,540	2,730			
22	Titanium (Ti)	4·507	1,670	3,260	11·58	4·14	0·34
23	Vanadium (V)	6·1	1,860	3,400	13·79	5·03	0·36
24	Chromium (Cr)	7·19	1,875	2,665	24·82		
25	Managanese (Mn)	7·43	1,245	2,150	15·86		
26	Iron (Fe)	7·87	1,536	3,000	19·65	7·93	0·28

Atomic number	Element	Density at 20°C (10³ kg/m³)	Melting point (°C)	Boiling point (°C)	Young's modulus (10¹⁰ N/m²)	Shear modulus (10¹⁰ N/m²)	Poisson's ratio
27	Cobalt (Co)	8·85	1,495	2,900	20·68	7·93	0·31
28	Nickel (Ni)	8·90	1,453	2,730	21·37		0·31
29	Copper (Cu)	8·96	1,083	2,600	12·41	4·62	0·35
30	Zinc (Zn)	7·13	419·5	906	9·65	3·45	0·35
31	Gallium (Ga)	5·91	29·8	2,240	0·97		
32	Germanium (Ge)	5·32	937	2,830	7·58		
33	Arsenic (As)	5·72		613ᵃ			
34	Selenium (Se)	4·79	217	685			
35	Bromine (Br)	3·12	−7·2	58			
36	Krypton (Kr)	0·0037	−157	−152			
37	Rubidium (Rb)	1·53	38·9	688	0·023		
38	Strontium (Sr)	2·60	768	1,380	1·72	0·69	0·28
39	Yttrium (Y)	4·47	1,510	3,030			
40	Zirconium (Zr)	6·49	1,852	3,580	9·65	3·45	0·34
41	Niobium (Nb) [Columbium (Cb)]	8·57	2,470	4,900	10·34	3·72	0·38
42	Molybdenum (Mo)	10·22	2,610	5,550	34·48		
43	Technetium (Tc)		(2,100)	(3,900)	40·68		
44	Ruthenium (Ru)	12·2	(2,500)	(4,900)	41·37	18·61	0·25
45	Rhodium (Rh)	12·44	1,965	4,500	28·96		
46	Palladium (Pd)	12·02	1,552	4,000	11·72	4·83	0·39
47	Silver (Ag)	10·49	960·8	1,761	7·58	2·76	0·38
48	Cadmium (Cd)	8·65	320·9	765	5·52	2·76	0·29
49	Indium (In)	7·31	156·2	2,000	1·10		
50	Tin (Sn)	7·298	231·9	2,270	4·69	1·72	0·36
51	Antimony (Sb)	6·62	630·5	1,380	7·58		
52	Tellurium (Te)	6·24	449·5	990	4·14		
53	Iodine (I)	4·94	113·7	183			

Atomic Number	Element	Density at 20°C (10^3 kg/m³)	Melting point (°C)	Boiling point (°C)	Young's modulus (10^{10} N/m²)	Shear modulus (10^{10} N/m²)	Poisson's ratio
54	Xenon (Xe)	0·0059	−112	−108			
55	Caesium (Cs)	1·903	28·7	690	1·24	0·17	
56	Barium (Ba)	3·5	714	1,640	6·90		
57	Lanthanum (La)	6·19	920	3,470			
58–71	(Rare earth elements)						
72	Hafnium (Hf)	13·09	2,250	5,400	13·79	3·45	0·37
73	Tantalum (Ta)	16·6	2,980	5,400	18·62	18·62	0·17
74	Tungsten (W)	19·3	3,410	5,900	34·48	15·17	0·26
75	Rhenium (Re)	21·04	3,170	5,900	48·27	20·69	0·26
76	Osmium (Os)	22·57	(3,000)	5,500	55·16	23·44	0·25
77	Iridium (Ir)	22·5	2,455	5,300	51·71	22·06	0·26
78	Platinum (Pt)	21·45	1,769	4,530	14·48	5·52	0·39
79	Gold (Au)	19·32	1,063	2,970	8·27	2·76	0·42
80	Mercury (Hg)	13·55	−38·6	357			
81	Thallium (Tl)	11·85	303	1,457	0·69	0·28	0·45
82	Lead (Pb)	11·36	327·4	1,725	1·79	0·55	0·45
83	Bismuth (Bi)	9·80	271·3	1,560	3·17		
84	Polonium (Po)		250				
85	Astatine (At)		(300)				
86	Radon (Rn)	0·01	(−70)	−61·8			
87	Francium (Fr)		(27)				
88	Radium (Ra)	5·0	700				
89	Actinium (Ac)		(1,000)				
90	Thorium (Th)	11·66	1,750	(3,850)	7·58	2·76	0·30
91	Protactinium (Pa)	15·4	(1,200)				
92	Uranium (U)	19·07	1,132	3,820	17·24	7·58	0·24

(Numbers in parenthesis are uncertain)   [a] Sublimes.

Answers to Problems

Chapter 1

1·2 820 eV

1·3 5.92×10^{-29} m

1·4 3·5 Å

1·6 166 Å

1·7 1.25×10^{14} cm^{-2}

1·8 2.5×10^{14} cm^{-2} s^{-1}

1·9 $y = \dfrac{\alpha}{4} - \dfrac{2\alpha}{\pi^2} (\cos \omega t + \dfrac{1}{3^2} \cos 3\omega t$

$\qquad + \dfrac{1}{5^2} \cos 5\omega t + \ldots) + \dfrac{\alpha}{\pi} (\sin \omega t$

$\qquad - \dfrac{1}{2} \sin 2\omega t + \dfrac{1}{3} \sin 3\omega t - \ldots)$

Chapter 2

2·1 8.64×10^9

2·2 5.2×10^4 ms^{-1}; 4.5×10^8 m^{-1}

Chapter 3

3·1 $3.36 \times 10^{-24} \, n^{2/3}$ J $(n = 1, 2, 3 \ldots)$

3·2 1·06 Å; 0·6 Å

3·3 -2.17×10^{-18} J; -5.42×10^{-19} J;
-2.41×10^{-19} J; 2.46×10^{15} Hz;
2.91×10^{15} Hz; 4.55×10^{14} Hz;
$n = 3$ to $n = 2$

3·4 5, 5, 7

3·5 $-e^4 \, \mathrm{m}/2h^2 \, \varepsilon_0^2 n^2$

3·9 0·1 eV

Chapter 5

5·1 10·6 eV

5·2 109° 28′

5·3 1.03×10^{-11} N, 2.32×10^{-10} N

5·6 1·33 eV, 4·8 eV

Chapter 6

6·1 $(\sqrt{3} - 1)$, 1·04 Å

6·4 9° 36′, 28° 6′

6·5 2, 4

6·7 2.25×10^3 kgm^{-3}

6·8 8.50×10^{28} m^{-3}, 8.50×10^{28} m^{-3}

6·10 1·83 Å, 1·95 Å, 1·27 Å, 1·44 Å

6·11 {111}

Chapter 7

7·1 17·9 cm^2

7·2 0·528

7·4 8.75×10^{-6}

Chapter 8

8·5 1,460 kgf/cm^2

Chapter 9

9·1 4.7×10^{24}, 2

9·2 8 km

9·3 3·61 cm

Chapter 10

10·1 (a) $\alpha = 42\%$, $\beta = 58\%$; (b) 0%;
(c) $\sim 85\%$;
(d) $\alpha = 37\%$; $\beta = 63\%$; (e) 76%

10·2 (a) 63%;
(b) Cristobalite 92·4%, Mullite 5·5%

10·3 (b) 100% γ; (c) see text;
(d) $\gamma = 49\%$, cementite 51%

Chapter 11

11·1 8·98 μm

11·2 2.6×10^7, 10^8

Chapter 12

12·2 8.5×10^{28} electrons/m^3,
7.4×10^{-7} ms^{-1}

12·4 4·09 eV, 2·48 eV, 8·54 eV, 6·76 eV

12·5 4.1×10^{-3} m^2 V^{-1} s^{-1};
6.7×10^{-3} m^2 V^{-1} s^{-1};
4.4×10^{-3} m^2 V^{-1} s^{-1};
9.0×10^{-4} m^2 V^{-1} s^{-1};
7.9×10^{-4} m^2 V^{-1} s^{-1}

12·6 2.3×10^{-14} s; 3.8×10^{-14} s;
2.5×10^{-14} s

12·7 1.58×10^6 ms^{-1}, 1.39×10^6 ms^{-1},
1.39×10^6 ms^{-1}, 3.68×10^{-8} m,
5.28×10^{-8} m, 3.48×10^{-8} m

12·8 $2.6 \times 10^{-8} \, \Omega$ m, 79 Å, 9·5 Å

Chapter 13

13·1 $0.58 \text{ m}^2 \text{V}^{-1}\text{s}^{-1}$; $\quad 0.26 \text{ m}^2 \text{V}^{-1}\text{s}^{-1}$;
$0.074 \text{ m}^2 \text{V}^{-1}\text{s}^{-1}$;
$2.84 \times 10^{26} \text{m}^2 \text{V}^{-1}\text{s}^{-1}$

13·2 $3.48 \times 10^7 \text{m}^{-3}$

13·3 9.47×10^{17} electrons/m^3;
2.19×10^{20} electrons/m^3; $\quad 0.48 \,\Omega\,\text{m}$;
$31.6 \,\Omega\,\text{m}$

13·4 1.77×10^{29} electrons/m^3; 1 in 2×10^{10}

13·5 $3.21 \,\Omega^{-1}\text{m}^{-1}$; 6.25×10^{18} donors/m^3

13·6 3.68×10^{-12} at %

13·7 $2,060\ ^\circ\text{K}$

13·8 0.51 mA

13·9 $1.37 \times 10^3 \,\Omega^{-1}\text{m}^{-1}$

13·10 1.06×10^{-4} W

13·11 1.57×10^{14} Hz

13·13 2.54×10^{-8} m

Chapter 14

14·2 3, 4, 1, 7, 2, 4, 3

14·3 1.75×10^6 Am^{-1}

14·4 4.97×10^5 Am^{-1}

14·5 6.21×10^5 Am^{-1}

14·6 1.63×10^5 Am^{-1}

14·7 7.6×10^{-5} Wb

Chapter 15

15·1 9.25×10^{-10} N; $\quad 1.16 \times 10^{-11}$ N

15·2 1.81×10^8 V

15·3 10^{-5} cm^{-2}

Index